T3-BFX-962

What's Causing
CREATIONS

Aliens
Evolution
God

Doug Hans

What's Causing Creations
Doug Hans, Author
Revised Edition 2018
NESNAH Publishing
authordoughans@gmail.com

Many thanks for permission from NASA to use their archive of non-copyright photographs on the cover and as illustrations in this book. Also, thanks to the brave men and women at NASA, and the many scientists through whose work makes possible the revelations of the wondrous Universe to the world.

This book may not be reproduced in whole or part, in any form or format, by any means including electronic, mechanic, photocopying, reproduction, or recording. The scanning, uploading, photographing, videotaping, and distribution of this book without the author's written permission is a theft of the author's intellectual property. Receipt of this book without a cover represents stolen property.

Disclaimer and terms of use: The author strived to be as accurate as possible in the writing of this book. The author does not warrant or represent that the fair use contents are accurate, due to rapidly changing theoretical information. All attempts were made to fact-check at the time of publication. The author assumes no responsibility for errors, omissions, and contrary opinions of the subject matter. Some of the contents of this book are the author's opinions. The author encourages readers to do their own fact-checking.

© 2018 Copyright holder author Doug Hans
Library of Congress Cataloging-in-Publication Data has been applied for.
ISBN 978-0-692-96648-8   Paper Back

"What's Causing Creations" has been recommended
for adults and young adults
as a well-written and stimulating
read on this topic.
Packed with the latest theories on aliens, evolution,
big bang, NDE, consciousness, anthropic principle,
abiogenesis and much more.

∞   ∞   ∞   ∞

"I'm recommending this book to all deep thinkers. It
explores theories in-depth and answers questions I
have had since a child."   Dr. Kellie Speciale

"I loved Chapter 10 (Life on Earth) -
so fascinating." Educator Karen Kiser

"What an eye-opener! Easy read with great lay-out
and presentation of facts."   Dr. Richard Rule

"This book is enlightening, thorough and convincing."
Dr. Larry Marcotte

This book is dedicated to those who are in awe of creations - creations of the Universe, Earth and of all living things that inhabit it.

Hopefully there will be more in awe after reading this book.

Introduction

Scientists say there are only three creation possibilities. Find out what evidence the intelligentsia offers that either aliens, evolution, or God created all that exists, beginning with the big bang and ending with all living things. The highly-educated intelligentsia influence the political, social, educational, and spiritual beliefs of their cultures. They include academia, scientists, theologians, writers, pundits, comedians, celebrities as well as others.

The spiritual account of creations is predictable, while scientific theories seem sexier and more appealing. Modernists believe everything that exists began from evolution. They also believe creationist believers have backward ideas, because it is by faith alone that they believe God created the Universe. Scientists are concerned that religions try to shelter young adults from the realities of scientific creations. Some philosophers and theoretical scientists argue that it is not possible to believe in an abstract concept of a creator God.

"What's Causing Creations" is packed with the latest creation information offered by the intelligentsia in simple terms, illustrations and definitions. It covers aliens, evolution, random events, God, abiogenesis, the anthropic principle, the big bang and much more. Find out what is being taught, including the difference between theoretical and provable.

World-wide streaming media makes new ideas instantly available outside classrooms, places of worship, and family institutions. Along with it comes new and sometimes unanswerable questions for the intelligentsia.

"What's Causing Creations" asks over 200 challenging questions about creation theories. Inquiring minds need to be better informed on *all* aspects of theories being proposed. Perhaps it is time someone asked questions.

Contents

Part I

Chapter 1   Only Three Creation Possibilities

Chapter 2   The Amazing Universe

Chapter 3   What the Intelligentsia Say

Chapter 4   Proof Means Different Things

            To Different People

Part II

Chapter 5   Did Aliens Create Humans

Chapter 6   Is Evolution Creation

Chapter 7   What Created the Big Bang

Chapter 8   What Created Atoms

Chapter 9   Why Is Earth So Different

Chapter 10   What Created Humans

Part III

Chapter 11   Did God Create

Chapter 12   Where is God

Part I

Chapter 1

Only Three Creation Possibilities

The Universe is magnificently amazing. Its vastness is beyond comprehension. From planet Earth, the Universe offers a showcase of galaxies, stars, planets, moons, meteors, asteroids, and comets. From outer space, even more splendor is seen. The Universe's wonderment is especially spectacular for those who pause for a moment to really understand and appreciate its complexities. Disinterested people are unaware and take for granted the Universe that is filled with creations, while others ask the big question – how was this enormous Universe created?

When scientists throughout the world debate what caused creations of the Universe and all that exists, the overwhelming consensus among them is that there are only three possible causes, and none others. It is either aliens, evolution, or God.

Only Three Creation Possibilities

- **Aliens**
- **Evolution**
- **God**

The dictionary defines creation as bringing something into existence out of a state of non-existence. But as only the dictionary can do, definitions of words are sometimes modified over time in ways that the meaning may be changed entirely.

A word that has been added to creation's definition is the word *produced*, such as humans producing a new painting. This opens the door to anyone becoming a creator, when in fact, there are only three possibilities for creators, and humans are not one of them.

Creation

**The act of bringing something into existence
out of a state of non-existence**

This book is about the Universe, Earth, and living things coming into existence where they did not exist before. In this book, *producing something* has nothing to do with *creations*.

People commonly use the word creation to describe something they made or produced. They did not create. Because something already existed, something cannot be *created* just because the shape or functionality of something changed. Builders may have produced a new house, others a new food recipe, or a new cancer-fighting drug, but none of these were creations. They existed before, only in a different state.

People often take credit for creating something they did not create. Instead, they use recycled atoms. No human has ever been able to create something out of nothing.

> **The word creation is often used inaccurately to describe something that is *produced* instead of *creating something out of nothing* - producing has nothing to do with creation.**

Atoms are the basic building blocks for everything that exists. Atoms have always existed, since the time of the big bang.

Creation is nothing like the magic act of pulling a rabbit out of a hat, which uses deception. True creation is something different. Creation is real magic because something is created out of nothing.

With only three possibilities for bringing something into existence out of nothing, then what created the three possible creators?

- **What created aliens out of nothing?**
- **What created evolution out of nothing?**
- **What created God out of nothing?**

Creations did not stop with the big bang. Even after this initial step, trillions of other creation steps were required, all culminating in the creation of all living

things including human beings.   Even today creations
continue.

```
13.8 Billion Years Ago
          \
    Science laws and systems
        Big Bang
          Universe
            Earth
              Protein molecules
                Single-celled living things
              Humans
                \
              Today
```

Major Steps in the Universe's Creation

Did one of the creation possibilities create other
possibilities?

- **Did aliens create evolution?**
- **Did aliens create God?**
- **Did evolution create aliens?**
- **Did evolution create God?**
- **Did God create aliens?**
- **Did God create evolution?**

It is almost an impossible task to defend any one of the three creation possibilities over the other two. Studying creations is an overwhelming task even for scientists who spend their careers trying to discover more about it. When they try to learn more about just a single creation step, they become extremely challenged, because creation is so complicated.

The most important creation step, creating something out of nothing, lacks answers. This probably explains why scientists focus on only small pieces of the creation puzzle during their life-long work. Most theologians avoid the scientific aspect of creations. They teach creations out of nothing, using familiar religious books that give no detail.

Aliens

Some scientists, reporting on behalf of their peers, say they do not accept aliens or God as the explanation for creations. Aliens, as creators, would most likely be considered pseudo-gods by scientists and unexplainable in scientific terms, just as God has been ruled the same in US courts.

What are aliens, and who created them? There is a great amount of speculation throughout the world about the existence of aliens. Does this high interest come from the alien's shocking looks and actions that have been perpetrated by media producers? Hour-long science channels obtain high ratings from viewers who watch stories of aliens living among humans undetected. Fictional novels about encounters with aliens and even love stories with aliens are flooding eBook internet retailers. Has the production of alien stories, using the latest digital technology, accelerated the acceptance of their existence?

Discussions of aliens is a topic that is well received all over the world. World-wide support groups insist aliens are real and that available proof is being suppressed by governments, especially their own. A high percentage of people throughout the world believe in aliens.

Belief in Aliens	
Atheists and agnostics	60 %
Islam	45 %
Jews	40 %
Hindus	35 %
Christians	30 %

US presidents, when asked, have publicly stated there is no evidence pointing to aliens and their existence. Even so, governments throughout the world as well as the

United Nations routinely fund listening devices to monitor evidence of digital or analog data being transmitted from outer space. There has never been a transmission, of any type, of an intelligent message in over 50 years of listening, causing many scientists to believe there are no aliens in existence. And yet, stories by science writers promise that soon communication with aliens will be a reality.

Some theoretical scientists are investigating meteors as a delivery agent for the transfer of microscopic chemicals called amino acids, as a seed to start life on planets like Earth. Should there be a study to investigate the true origin of these meteors?

Evolution

What is evolution? Who created it? These may sound like silly questions, but if evolution is a creator agent, shouldn't more be known about it than just accepting it as a fact?

People do not seem to understand what evolution means. Evolution has to do with how species evolve and make changes and adapt generation after generation. It has been assumed, after hearing that evolution equates to creation, that must be what the definition is. Not so.

Evolution

**A *change* in, not the *creation* of,
the inherited characteristics of a
single specie through many generations**

Scientists have picked evolution as the creator agent for all things in the Universe. This creator agent would need to rely solely on *random events* happening to achieve its goals, if indeed it had goals.

Presumably scientists use the term evolution as a simpler way of communicating science theories about creations to the general public. In almost all discussions by theoretical scientists, evolution is exclusively used as the explanation of how everything, starting with the big bang and ending up with all living things, was created.

Scientists use the terminology
evolution
(not *random events*)
as a simpler way of communicating
science theories about *creations*

Evolution will be covered extensively in Chapter 5.

From the time of the theoretical big bang to the culmination of human beings on planet Earth, major creation milestones took place over billions of years. In order to focus on the major creation steps, trillions of steps in between the major milestones are not included in this book. However, these trillions of creation steps are continually happening to this day.

The Universe is a result of a continuum of gigantic complex creation steps that no one fully understands.

God

Who is God, and who created God? No one, including theologians, have answers to these questions.

Theologians believe God is the creator of the Universe based on the account of creation events in the Bible, Torah, and Quran. Few theologians go past the contents of the Bible to explain creations. Instead, they rely on faith for their belief in how creations occurred. Almost all theologians do not stress the importance of the creations in the past or that are continually happening today.

Many scientists are adamant that theologians have no proof there is a God, and that they rely on faith alone to explain creations. Some scientists believe explaining creations, using faith alone, is troublesome for societies because it does not allow people to understand how scientific events created the Universe. Should religions teach more about science and creations? Does it contradict the Bible or contradict the interpretations that some religious leaders have established?

**Do scientist's findings
contradict the Bible
or contradict the interpretations
that some theologians have established?**

Some theoretical scientists believe there is no God. They insist that the creations of the Universe can be explained by science alone. They believe God is an unprovable theory.

One billion people on Earth, out of the total seven billion people, have never been introduced to the concept of a god. For those who do believe, some do not believe their god or gods are the creator.

Europe is moving away from a belief in God towards a version of religious spirituality called spirit-force. It is not clear if this is considered a religion because it is radically different to the religions practiced in Europe in the preceding 500 years. One thing for certain is God appears abandoned in growing numbers as landmark European churches are increasingly phased-out for new uses, such as pubs.

Spirit-force has a growing fan base, but the concept seems to lack a coherent definition of what it exactly means. It could be driven by a culture or perhaps a movie that created the movement's beginnings. In general, it is thought not to be a person or deity. Rather, it is thought to be a force, energy, or power that may reside either in the living, the dead, or for that matter, in the existence of everything. It is also thought by some to involve energy moving from the dead to the living.

	God	Spirit Force
Greece	80 %	15 %
Italy	75 %	15 %
Germany	50 %	25 %
UK	40 %	40 %
France	35 %	25 %
Sweden	25 %	55 %

Are Movies Introducing New Creators?

The Star Wars movie talks about a galaxy far away from Earth where no god exists. Characters in the movie substitute phrases like "may the force be with you", instead of the more common phrase "may God be with you".

In the movie Avatar, a spirit god called Eywa presides over the tree of souls on the grounds of Pandora, located on the floating Hallelujah mountains. In this movie, trees and animals have ways to connect directly to humans by way of their nervous systems.

Are these movies, perhaps, depicting a substitute for one of the three creation possibilities?

If science is correct that there are only three possibilities for the creations of the Universe, aliens, evolution or God, which one is it? Scientists say it is evolution. Theologians say it is God.

Chapter 2

The Amazing Universe

Scientists are studying the vast Universe with its undefined borders. They have documented galaxies, stars, moons, comets, and meteors and published theories about the Universe's beginning, how fast it is moving, and when it will end. Theologians explain biblically what happened when the Earth began but their explanations offer a stark contrast to that of theoretical scientists' explanations.

Is the Universe a Cosmos?

Does planet Earth reside in a universe or a cosmos? The collection of stars and planets, seen and unseen, are all part of either a universe or a cosmos. Both mean pretty much the same thing, except for one big difference in their definitions.

The term universe is defined as a complex and ever-changing system, full of chaos. A cosmos is defined the same as a universe, but is orderly, not chaotic.

Universe

**A complex and everchanging
system of stars and planets
with chaos**

Some people think Earth is part of a cosmos. Because the cosmos is most commonly called a universe, in this book universe will be used, even though cosmos may describe Earth's presence in the Universe more correctly.

Cosmos

**A complex and everchanging
system of stars and planets
without chaos**

Are There More Universes in Outer Space?

In addition to the theory that the big bang created the Universe, there are alternate theories about the creation of other universes. These theories include: *string theory*, *M-theory*, and *multiverse theory*, to name a few. At first these theories were strongly rejected by big bang theoretical scientists but now are accepted as real possibilities.

Metaverse is a new term starting to be used but is not an alternative universe theory. Instead, it is a virtual reality space that combines world experiences and human brain knowledge into common clouds for public domain sharing. Its proponents say virtual reality space will broaden mankind's encounter capabilities by using artificial intelligence, time irrelevance, and human consciousness expansion after death.

Will this artificial intelligence enable a human's consciousness to continue after death as the developers of virtual reality intend?

Computer scientists are continually developing software to engage the human mind with spatial trickery. Virtual reality, augmented reality and 7D holograms are three of their recent products.

Virtual reality is technology using realistic 360-degree viewing, placing a person's mind at another person's location, any place in the world. Could it be used sometime in the future to visit anywhere in the Universe?

Augmented reality occurs at a person's location while interjecting foreign objects into a person's viewing and senses. Both employ headgear that are modified eye glasses, which superimpose images and information onto the lenses. During 2018 the first commercial offering of prescription eyeglasses with augmented reality was introduced using informational display as a side benefit to ordinary reading glasses.

Seven-D hologram reality uses split laser focusing of a recorded image onto a reflective surface creating a light field. It adds four additional attributes to each of the x-y-z pixelization of such an image providing for such things as movement and sensations over an ordinary stationary image.

These technologies will ultimately be used by marketers of products or ideas, including the possibility of bringing the deceased and "heaven" into a living person's reality. Where will this technology eventually lead mankind?

Scientists say the Universe is expanding. Because it is expanding, and its boundaries are undefined due to the bending of light waves, it is unlikely parts of the Universe will ever be explored by mankind, even when using high powered telescopes. People will have to settle for scientists' theories as they wonder what is at the outer edges of the Universe.

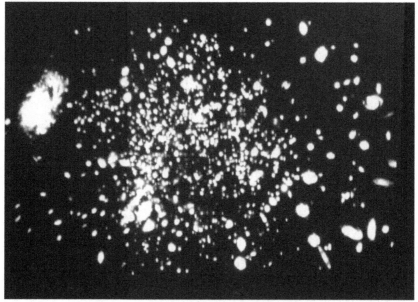

No One Knows Where the Edge of the Universe Lies

Galaxies Are Flat and Spinning

Scientists say the Universe, galaxies, and Solar Systems all spin like pinwheels around some central point. They are also relatively flat, like giant platters.

If the big bang explosion began at some central point and accelerated outward in all directions, then why are galaxies flat? Was there a celestial retainer wall that stopped the expansion of particles from spreading outward into all directions?

What started the spinning of these giant platters? Theoretical scientists say it takes galaxies 200 million *light* years to make one complete rotation. Were there celestial trade winds, like the trade winds experienced on planet Earth, that directed atoms to start the spinning of the galaxies? Scientists say that after the big bang, empty space was filled with atomic dust, which clumped together and somehow started the spinning.

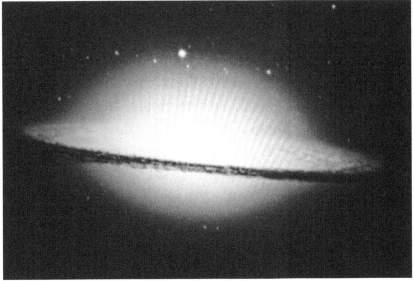

Galaxies Are Flat

Eyes in the Sky

The Hubble telescope, operated by NASA, was launched in 1990. Its discoveries have opened mankind's Earth-bound eyes to the immensity of the Universe. The telescope sends back amazing photographs and research data of nearby galaxies like the Milky Way, where planet Earth resides. By the year 2019, a new rocket will be orbiting around the sun which will launch a new telescope the size of a 15,000 square foot three-story building. The new telescope, called the Webb, will be 100 times more powerful than the Hubble telescope. Can this be imagined? How can it beat the amazing photographs that the Hubble telescope has already taken?

Because the new telescope uses infrared wavelength technology, it will be able to look backward in time 13 billion years earlier, taking it to nearly the time of the theoretical big bang. Equally important, it will look deeper into space revealing more about how the Universe's stars and planets were formed. This represents a revolutionary new tool for scientists, enabling them to theorize in more detail the steps involved in the Solar System's creations.

Timeline of the Universe

The timeline of the Universe is continually being mapped. Theoretical cosmologists and physicists calculate the origins of the Universe at something over 13 billion years ago, with planet Earth about 4.5 billion years ago.

Theoretical Provable

?	13.8 billion	4.5 billion	300 million	2.8 million	9,000 BC	2,500 BC	2,000 BC	1,500 BC	800 BC	700 BC	50 BC	59 BC	65 BC	80 AD	93 AD	1608 AD	1665 AD	1939 AD	1948 AD	1990 AD
Science laws and systems created	Big Bang Universe created	Earth, protein and bacteria molecules created														Telescope	Calculus		Carbon dating	Hubble telescope / rocket
Space	Gravity	Energy and matter	Stars and planets					Carbon-dated fossils								Theoretical				Historians
	Atoms															Theoretical physicists and cosmologists			paleontologists, archaeologists, biologists	
					Telescope Calculus Rockets															

Timetable of the Universe

Mankind began recording history 4500 years ago, thanks to the Egyptian's discovery of papyrus and ink. Drawings on cave walls earlier than 4500 years ago helped to piece together additional history. These limited drawings however, only go back in time to the very near past of the Universe's history. Looking at the Universe's timeline, it becomes apparent that humans fit into just a tiny slice of the total time of the Universe's existence.

Creation events can only be empirically proven that happened in the last few thousands of years of the Universe's history, but they still seem likely to have happened.

Empirical

**Something that is verifiable
by observation or experience
rather than theory or logic**

With the invention of calculus, carbon dating, and rocket-launched telescopes, it is now possible to more clearly calculate and understand planet Earth's development within the timeline of the Universe.

Most of what is known about events on the timetable are theoretical and may never be proven empirically because of limited observation, time constraints, and distance. Regardless, theoretical scientists have done an amazing job trying to piece together the timeline of the Universe.

Anthropologists, archaeologists, biologists, and paleontologists use a tool developed for them called *carbon dating*. This tool examines dead tissue from a plant or animal that was living sometime in the past. It compares dead tissue to living tissue and determines how much time has taken place since its death.

Carbon dating can accurately predict when a death of a living thing occurred, plus or minus 100 years of its death. When looking back beyond 50,000 years or so, carbon dating's precision has limited accuracy. Occasionally, samples can be dated further back, under special conditions.

The Solar System

Earth's star, or commonly known as the Sun, is tremendously hot with an estimated temperature of 28 million degrees Fahrenheit. Nuclear fusion powers the Sun by converting hydrogen atoms to helium atoms. This provides Earth with a tremendous source of heat, light, and energy. The sun's surface is violent, and at times showers Earth with solar winds. This violence causes explosions that have the capability of damaging electrical devices such as computers and power grids.

Fortunately, an iron core inside Earth's center provides a strong magnetic shield. This propels most of the Sun's solar particles from direct hits as the solar winds curve around Earth. The byproduct of this event is a fantastic light show at the north and south poles, known as the aurora borealis.

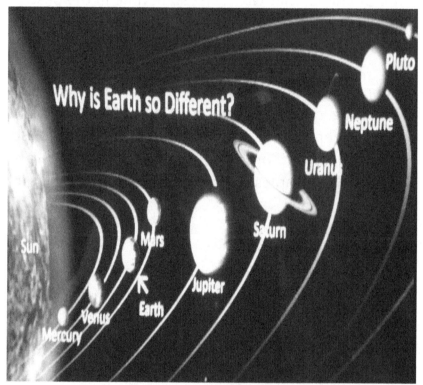

Planets in Earth's Solar System

Without the Sun's energy, the Earth's rotation, and atmosphere, planet Earth would be either too cold or too hot for living things to exist.

Why is Earth Unique?

A big question that has puzzled scientists for centuries is why is planet Earth so different from the other planets in Earth's Solar System? As an example, why did Earth receive so much water? Without water, life on Earth could not survive. Other planets created in the same Solar System, in the same habitable zone, did not do as well. Why didn't some of the other planets receive sufficient supplies of water molecules to support life like planet Earth?

Theoretical scientists are continually looking for other planets in the Universe that are possibilities for supporting life. Because of the Universe's vastness, scientists are only looking in Earth's Milky Way galaxy. The Milky Way is immense all by itself. Scientists are convinced there are other habitable zones besides Earth's that are in other Solar Systems in the Milky Way. Not much is theorized about life in other galaxies other than the Milky Way, because they are too far away to study.

Earth's Crust is Fragile

Earth is covered with a thin crusty surface. Below the 25-mile thick surface is molten lava that is a fiery 10,800 degrees Fahrenheit. Almost everyone on this planet takes this for granted, while going about his or her daily life. Because Earth is red hot under the surface, it is probably a good thing people do not think about it too much, especially when visiting such thin-crusted places as Yellowstone Park, known as the largest volcano in the world.

Everyone on Earth's ever-changing and shallow surface can observe physical changes as Earth redefines itself. Massive amounts of dust and sand are blown or washed away each year. Wind storms, fire, lightning, tidal waves, earthquakes, and volcanic eruptions are always changing the dimensions of mankind's living space in Earth's habitable zone. Most animals and plants make the necessary adjustments for survival.

A billion years ago, almost all living things could not have survived because Earth was so volatile. A billion years from now no one knows if life will exist on this planet. Theoretical scientists are most likely already contemplating these possibilities.

The Habitable Zone

Most theoretical scientists believe Earth's star was formed from cold, swirling atomic particles when the center collapsed to form the Sun. Millions of years later scientists theorize hot chunks of atoms peeled-off to form planets.

Planets in the Sun's Solar System vary in temperature from minus 400 degrees Fahrenheit on planet Pluto to plus 900 degrees Fahrenheit on planet Venus, depending on their distance from the Sun. Fortunately, Earth lies within a habitable zone, or correct distance from the Sun, which provides a sustainable atmosphere for a wide variety of living things.

Most of the Sun's heat is directed toward the middle region of Earth's surface. Planets that were created at the right distance from the Sun, with the correct rotational pivoting axis, provide a habitable zone for living things to exist. The predictive wind patterns and rotation of Earth cause heat to be distributed widely, while being retained under clouds. This is fortunate for the survival of animals and plants.

Earth's inhabitants can thank planet Jupiter for their relative safety from meteor collisions. With Jupiter's large size and stronger gravitational force, it helps to shield Earth from possible deadly asteroids.

Are Humans the Only Inhabitants of the Universe?

Could it be that humans, residing on Earth, are the only humans or humanoids in this vast Universe? No one on planet Earth knows for certain.

The diameter of the Universe is estimated to be 90 billion light years across, and it is growing. There are 200 billion galaxies in the Universe.

Scientists estimate the Earth's Milky Way contains about 400 billion stars, with about 100 billion planets. Due to the limitations of time, scientists can only speculate on the possible life that may reside on other planets within the Milky Way galaxy. Scientists are not studying possible life outside the Milky Way galaxy.

The Milky Way Galaxy

Theoretical scientists have estimated there are 11 billion planets rotating in habitable zones around stars, like Earth's Sun. Again, these are all inside the Milky Way galaxy that is about 100,000 light years in diameter.

The nearest galaxy to the Milky Way is 163,000 light years away. If it were possible for someone to visit the nearest galaxy to Earth, they would have to travel at the speed of light, or at an equivalent highway speed of 670,000,000 miles per hour (MPH), for 163,000 years.

That's years!

Miles per hour is calculated by multiplying the speed of light, which is 186,000 miles per second, by 3,600 seconds per hour. Technology has not been developed to travel this fast.

A light year's distance equals traveling at 670,000,000 miles per hour . . .
for one year

Earth is 27,000 light years from the center of the Milky Way. This makes it about half-way between the middle and the outside edge of the galaxy.

Scientists have discovered several planets that are theoretically like Earth, but in other Solar Systems. At least these planets reside within the same Milky Way galaxy as Earth's solar system which make them possibilities for space exploration.

Because of the limits of present technology, no other planet like Earth has ever been observed closely by either a telescope or sound antenna.

How many star gazers realize the depth of the Universe while looking up on a clear night? Remember, stars seen with the naked eye are only the stars of the Milky Way, which is just one galaxy out of the 200 billion galaxies in the Universe.

Two of the Billions of Galaxies in the Universe

Scientists say the Universe is an immense, boundless space that is expanding rapidly. Most of the Universe is empty space. *Atoms*, the basic building blocks of the Universe, occupy only about one percent of space inside the boundaries of the Universe, while energy and dark matter occupy the remainder. Some scientists today theorize that dark matter may not even be matter, but instead another form of gravity.

Is it any wonder why some people find it hard to imagine that there could be another life-sustaining planet like Earth out there in the stark vastness of the Universe? Others couldn't disagree more with the doubters as they speculate about possible far-away outer-space relatives.

Where are the Universe's borders? Where does the Universe begin, and where does it end? Because of time and the bending of light waves, some theoretical scientists say the outside edge of the Universe will never be observed because it is always over the horizon.

Scientists are continually discovering new stars and galaxies as the result of nuclear explosions within the Universe. They will undoubtedly discover more and more about the life cycle of stars and planets as newly designed space telescopes are positioned in the skies.

In just a relatively few short years, scientists' exploration has opened-up the skies for mankind to view what was previously unknown. Powerful celestial software Apps now allow Earth's people, through simulated viewing, vivid detail of the many splendors of the Milky Way galaxy. Will these Apps soon be able to go beyond?

Do stars have a life cycle of birth and death? Are there other planets like Earth out there? Are there other humans, plants, and animals somewhere in the Universe?

What wonderment in the Universe is yet to be discovered?

Chapter 3

What the Intelligentsia Say

What do the intelligentsia say about creation? Who are the intelligentsia? They are a well-educated, elite class who influence the political, social, scientific, educational, and spiritual culture of their societies. They include leaders from governments, academia, scientists, theologians, writers, filmmakers, celebrities, athletes, comedians, and even pundits.

Their positions allow them to express their beliefs to the general population. This is not necessarily a negative thing, merely how the world currently operates. Most people follow the ideas of the intelligentsia, and over time, their ideas are accepted as fact. Some leaders' ideas are followed but they are not considered intelligentsia because they obtained their positions through brute force.

A large portion of the world's population consist of those who are challenged just to stay alive. They may or may not feel the influence of their intelligentsia today, but in the future their belief systems could change rapidly as they are electronically invaded with different points of view from different parts of the world. For those who live under a dictatorship, information channels may be blocked.

Technology and the Local Intelligentsia

In the next ten years, digital media will be within an earshot of most people on this planet. In 2017, 3.8 billion people around the world watched the world cup soccer games from a multitude of video devices. Future messaging will deploy this kind of technology and will stream virtual reality video from its original source to far away countries.

The last sixty years have brought an unparalleled explosion of communication devices. Whole generations are left behind, while new generations can't get enough of the new technology.

World Communication		
	Devices or Users	First Introduced
Land line phones	1.0 billion	1876
TVs	1.4 billion	1946
Cell phones	6.8 billion	1973
Internet users	3.3 billion	1986
Streaming video	.4 billion	1992

No one can predict what the new technology will be or its effect on spreading new ideas throughout the world. Will streaming video or other technologies present challenges for the intelligentsia?

Prosperity in the World

Household income will rise rapidly in the world, causing a rippling effect on local economies, technology, and education. Third world countries may outpace countries like the US in economic growth. This could cause local unrest and questioning of beliefs given by their intelligentsia.

Years it Took to Double Household Income

China	12 years
Vietnam	17 years
India	19 years
Indonesia	26 years
US	110 years
UK	195 years
Italy	455 years

As disposable income increases so does technology, access to information, trade, education, and wealth.

Household Income Projection

	2010	2050
Switzerland	$38,700	$83,600
Hong Kong	$35,200	$76,200
Japan	$39,400	$63,300
US	$36,300	$55,700
Germany	$25,100	$52,700
UK	$27,600	$49,400
South Korea	$16,500	$46,700
Italy	$18,700	$38,400
Vietnam	$1,300	$20,000
China	$2,400	$17,400
Indonesia	$1,200	$5,200
India	$800	$5,100

In US Currency

Good paying jobs will be the driver, while household incomes rise rapidly over the next 40 years.

World Population

World population is estimated to continue rising over the next 40 years, mainly in India, Indonesia, and the US. China controls their projected rise in population with a mandatory abortion law.

	Population (millions)		
	2011	**2050**	
China	1,337	1,304	-2%
India	1,190	1,650	39%
Indonesia	246	313	27%
US	313	429	40%
UK	61	65	1%
Italy	58	55	-1%
World	7,000	9,256	+34%

Every second in the world four babies are born. Every second in the world two humans die. Will there be enough food to feed the new people? Will there be enough air to breathe and water to drink? Will there be enough land, land-fills, hospitals, and cemeteries?

In the next 40 years with a 34 percent increase in population in the world, *will there be enough available atoms* to create the bodies of the additional people?

**Will there be enough atoms
to create the two billion additional humans
on Earth by the year 2050?**

Ethnicity Shifts in the US

Ethnicity may take on greater importance in countries throughout the world, rather than race or other factors as cultures and races assimilate into one within regions. Technology may remove regional ideological boundaries.

Keeping this in mind, the next 40 years will see dramatic shifts in ethnicity in the US and other countries, just like it did in the last 50 years. Undoubtedly, other countries with porous borders will experience the same.

Ethnicity Shift in the US			
	1950	2005	2050
White	85%	67%	47%
Black	11%	13%	13%
Asian		5%	10%
Latino	4%	15%	30%

The most striking predictions for a country such as the US will be a change from a white majority to a white minority, while remaining a plurality of the population.

Age Shift in the US

The shifting of age groups in the US will bring dramatic pressure on the delivery of medical and economic stability. The tug-of-war between automation, higher paying jobs, and family survival will heighten, causing turmoil in cultures as politicians try to cope.

Age Shift in the US			
	1950	2011	2050
0-19	34%	27%	25%
20-24	7%	7%	7%
25-64	50%	50%	41%
65-79	8%	13%	20%
80+	1%	3%	7%

Will all the changes in demographics coming in the next 40 years, coupled with an expanding array of new ideas and theories, shift the current intelligentsia's thinking to embrace new ideas and the use of new technologies to help them better cope with change?

Intelligentsia Views on Creations

Today theoretical science is a dominant, educational force. As the public becomes more educated, not only in classrooms, but also from watching streaming educational videos, the views expressed by their current intelligentsia may become challenged.

Theoretical science is where the money is right now. Federal and private investment dollars encourage higher education institutions to pursue theoretical science. Most university science professors understand this. It is hard to find a large or small college that does not have, or is not in the process, of budgeting for a theoretical research park.

Lawsuits from non-science colleges are increasing, as professors are required to apply for grants to cover their salaries while academic dollars are spent elsewhere on theoretical science.

Some professors say that the greatest conflict of the 21st century will not be between the West and terrorism. Rather, it will be between a modern civilization and anti-modernists. They say anti-modernists are religious people who are taught what to believe. They are taught that they owe their allegiance to a higher order and believe they should prepare for an afterlife. Modernists, on the other hand, believe in science, reason, and logical truths as the basis for their beliefs.

Some scientists say there is no God or intelligent designer involvement in creations. Instead, they say Earth spontaneously created itself out of nothing through random events or quantum fluctuations. They further state that there is no evidence that planet Earth was visited by aliens. Scientists acknowledge they do not know how life first began.

- **Earth spontaneously created itself out of nothing by random events**
- **There is no intelligent designer**
- **Do not know how life first appeared**
- **The evidence shows we were not visited by aliens**

Some theoretical scientists say religions should embrace science because, if properly understood, science is so much more exciting, poetic, and filled with sheer wonder than anything in the poverty-stricken arsenals of the religious imagination.

Other scientists say that religion is bad for people because it teaches them to be satisfied with no scientific explanation of creations and allows them to believe creations happened by faith alone. Some scientists believe questions from religious people are easy to answer because their religions cannot prove God exists.

Scientists say religion is bad for people because it teaches them to be satisfied with no scientific explanation of creations and allows them to believe creations happened by faith alone

---

Some scientists assure religious people looking for answers about creations that scientists are close to providing proof there is no such thing as a God creator. These same scientists believe they should have leading roles in education and government because they are the intelligentsia.

What some scientists say about religious followers:

- **Religious people cannot prove God exists**

- **Science is close to proving there is no God**
- **Scientists are the intelligentsia and should have leading positions in government and education**

Other theoretical cosmologists in major universities teach students they do not need a deity to understand creations. They say one only needs to understand that quantum fluctuations are the creator of the Universe.

You don't need a deity to understand creations.

**You only need quantum fluctuations
to create the Universe.**

Some of the religions in the world attribute creations to a spiritual God. Almost all religions do not explain, even at a basic level, what creations are about.

Some of the religious intelligentsia have views about what caused creations, but they are not specific on how creation works. Scientists are more specific. Both groups meet the definition of the intelligentsia for their cultures as they teach their perspective views.

Regional religions are formed throughout the world from the beliefs of only a few intelligentsia leaders. This may radically change as digital media becomes independent of territorial boundaries.

As for scientists, their explanation of creations is thorough but theoretical at best. They do not explain creations in empirical terms because creations cannot be proven. They theorize about some of the creation events using such words as "might", "could", and "probably" to explain pieces of the creation puzzle.

Some scientists and philosophers who have non-theistic views on God, gods, or absence of a belief in a god, believe religious people create gods in their image. These people believe:

- **Science is the sole player in the natural world while religion's sole role is answering questions about ultimate meaning and moral value**
- **God is a symbol of human aspirations and values**
- **Religions construct human features into gods and spirits to make them more human-like, social, and familiar**
- **The existence of God is an empirical question which must start out by proving who created God**

The three largest Abrahamic religions in the world, Judaism, Christianity, and Islamism trace their origins to the practices and the worship of the God of Abraham.

The two largest religions, Christianity and Islam, comprise almost half of Earth's population. Neither religion seems to address creations in a way for those who have curiosity beyond a biblical knowledge to become thoroughly convinced.

The Christian religion is two thousand years old. The Islam religion is 1500 years old. Christians number 2.1 billion people while Muslims make up 1.6 billion. Another 1.5 billion are agnostic, 950 million are Hindus, and 385 million are Buddhists.

There are 13 world religions found mainly in China that comprise another 347 million of the world's population. The Jewish religion, one of the earliest religions, has 14 million followers. The Shinto religion in Japan has four million followers.

In the Christian religion alone, there are over 33,000 different denominations. In the Islam religion, there are at least three major sects and dozens of divisions within each sect. Multiple divisions within Buddhism, Hinduism, and Judaism also exist. These divisions are all driven by intelligentsia faith leaders with different viewpoints.

No wonder with so many different versions of religions in the world it is impossible for many people seeking spiritual faith to know where to begin. They often rely on their regional intelligentsia in their search for answers. Will they in the future? Because some older religions believe in multiple gods and other religions do not believe in any gods at all, trying to understand religion is even more complicated. Young people in free societies either drop out or explore far different versions of religions than they may have been introduced to earlier in their lives.

Religious Beliefs		
Christianity	2,100,000,000	30%
Islam	1,600,000,000	23%
No religion	1,500,000,000	22%
Hinduism	950,000,000	14%
Buddhism	385,000,000	6%
14 other religions	351,000,000	5%
Judaism	14,000,000	
World population	7,500,000,000	

In large countries such as China, there are old religions, seemingly buried in the landscape that have been around for centuries. No less than 150 million people are informal members of just one religion called the Old Chinese religion. In remote areas of the world, many tribes do not have the slightest idea what religion is or have a concept of what a religious structure for worshiping looks like.

Hinduism began about 2000 BC while Buddhism began around 600 BC. Neither religion focuses on a central god or offers an explanation on how creations happened. Judaism also began about 2000 BC but focuses on one God as a creator. These religions began centuries before the religion of Christianity was started. Do the religious intelligentsia, in executing their individual leadership in establishing religious doctrine and maintaining their organizations unintentionally drive people apart?

> # Do the religious intelligentsia unintentionally drive people apart?

Forty percent of the people in the United States are called Young Earth Creationists because they believe the Universe was *created in the last ten thousand years*. Most theologians who have listened to theories on Young Earth versus Old Earth admit they do not know how to decide between the two and that they certainly could not debate the subject. Because there is such a diversity of religions with diverse beliefs on creations, many young people seeking answers today receive limited guidance from theologians and are often confused.

Orthodox Jews tend to believe the creation account in the Torah while other Jewish followers have mixed views. Some have strong beliefs in evolution while some believe in both.

The last two Popes are not silent in their views on creations or evolution. Like so many other religious leaders around the world, they do not explain how creations happen in scientific terms. Instead, they introduce God's involvement into scientific theories that have already been presented.

The Roman Catholic Church once forbade their followers to read Darwin's evolution book. But now they profess that evolution is true, and God is its real author. They say scientific theories relating to the origin of the Universe are not in conflict with faith. They also profess that God was behind the big bang theory.

The following are thoughts from prominent intelligentsia theologians who appear regularly on television.

- **The biblical account of creation in Genesis matches the big bang theory**
- **The span of time between the first two verses of Genesis is the Gap theory, which puts billions of years into Genesis**
- **The current theory points to a big bang as the beginning of creation about 15 billion years ago**
- **God created man, whether it came by an evolutionary process or not. God took this being and made him a soul, which does not change the fact that God did create man**

These intelligentsia are iconic religious figures from every major religious organization in the US. Their beliefs are bold statements people most likely do not associate as coming from them while at the pulpit.

They say the big bang theory and evolution in nature do not contradict the notion of divine creations because evolution first requires the divine creations of the living beings that evolved. They see no reason why God would not have used a natural evolutionary process in forming the human species. Further, they state that God created all living beings, and God let them develop in accordance with the internal laws God gives to each one.

Another theologian said a person's birth at a geographic location determines to what religion a person belongs. This theologian says that those who are born in Pakistan are Muslim, in India they are Hindu, in Japan they are Shinto, and, if born in Italy, they are Christian. He says people naturally believe in exclusiveness, when it comes to their religion and their individual faith. Will this be challenged by streaming technology?

People naturally believe in exclusiveness when it comes to their religion.

Will this be challenged by streaming technology?

Some scientists believe Earth was created 4.5 billion years ago while some religions believe that Earth was created 10,000 years ago. In 1450 BC, Moses wrote in his book of Genesis a description of the creation of the Universe, including Earth over a *six-day period*. Ten years later in 1440 BC in Moses' book of Psalms 90, he also wrote that a thousand years to God is like only a few hours to man. Based on this, Moses knew Earth could be at least *billions of years* old, the same as scientists have concluded.

Why is the science side of creations not talked about more by the theologian intelligentsia? Aren't creations one of the major cornerstones of religion?

Buddhism rejects the existence of a creator deity. Instead, they believe that all focus should be on the suffering of mankind.

Hindus believe in multiple gods and that none of their gods are responsible for creations. They believe in reincarnation. If you ask two different Hindu followers what they believe, you are likely to get multiple explanations about their religion and its meaning. Hindus believe the world is an illusion.

The Christian, Judeo, and Islamic religions explain creations in their books of Genesis and Surah, but not the unbelievable creations' events in detail. Scientific theories on creations offer a more detailed account than the theologians' explanation.

Recent polls indicate theologians have not done enough to convince their followers that God even exists. There also seems to be a disconnect between people and God as their creator.

Religious inquiring minds are left with only the first two chapters of Genesis to explain creation events compared to volumes of books with current theories presented as facts in schools by the scientific intelligentsia.

For adults in the US under 30 years of age, only *fifty percent* believe in God with a high degree of confidence. Only *seventy-five percent* who call themselves Christians can say for certain that God exists. Only *thirty-three percent* of Jews, Buddhists, and Hindus say for certain God or a universal spirit exists.

Only 75 percent of Christians are certain that God exists. Only 33 percent of Jews, Buddhists, and Hindus can say for certain God or a universal spirit exists.

Could it be by not teaching a little more about *scientific creations*, theologians have minimized an important religious cornerstone and, unknowingly, passed it into the hands of science's intelligentsia to explain?

Are religious followers ready to have their intelligentsia explain scientific creations as part of their faith? Do religious followers understand the scientific fact that their bodies are being *re-created* every 3.4 years? (See Chapter 10).

Chapter 4

Proof Means Different Things To Different People

There are as many opinions in the world, it seems, as there are people on the planet. Trying to get consensus at a public or private meeting is sometimes impossible to do because of varied opinions. There are no exceptions to this when it comes to opinions on *What's Causing Creations*.

What is Proof?

It may sound like another silly question, but what is real proof all about? Obviously, if someone sees, feels, or touches something tangible, most people accept this as proof. But not necessarily all. Proof is only accepted on an individual basis and usually require compelling evidence.

Proof is a personal thing. It means something different to each individual person. Sometimes people are reluctantly forced to follow someone else's definition of proof. This, however, does not mean they believe something they are forced to believe. On the other hand, a great deal of people are followers of others and their ideas. They will not challenge the ideas of others, especially if the ideas come from people in positions of authority.

Proof

**Evidence of something
that compels acceptance
on an *individual* basis**

Bees in a hive will not accept the location of flowers from scout bees returning if the scout bees do not have nectar. They need proof. When it comes to humans, few ask other humans for guidance. People prefer their own proof. Often, they discover answers for themselves, even if multiple mistakes are made along their learning path. This is where the intelligentsia show up. They view their job as giving out answers. It doesn't matter whether the intelligentsia is a philosopher, theoretical scientist, teacher, politician, journalist, professor, government official, theologian, celebrity, pundit, or comedian. One of the self-proclaimed missions of the intelligentsia is to tell people what to believe, whether it is considered right or wrong by the majority.

People Get Answers from the Intelligentsia

People get answers from the intelligentsia that sometimes stay with them a lifetime. Believing answers, with or without proof, depends on an individual's circumstances, background, and how they are wired to accept someone else's answers. Some people are so hard-wired they lock in what they believe at a certain age and seldom accept another person's answers or ideas. People get answers to their questions in many ways:

Intelligentsia	Followers
Dictators pick the answers	Reluctant
Professors pick the answers	Unchallengeable
Journalists pick the answers	Pravda
Politicians pick the answers	Good Luck
Bafflers pick the answers	Gruber
Theologians pick the answers	Faith
Theorists pick the answers	Dazzled
Philosophers pick the answers	Null
Peers pick the answers	Millennium
Parents pick the answers	Respectful

People Can Be Fooled

Accepting another person's ideas is a mental process that occurs in several ways. There are ideas that are so difficult to understand that people succumb to accepting whatever highly-credentialed people tell them to believe.

At other times, by adding a little mischief called baffling, an idea can be pushed over the top, like purchasing an ice cream cone because chocolate has been added on top for free. It just makes a person want it so much more.

Baffle
To defeat someone by confusing

People Want to Believe

When it comes to science, most people believe theoretical information presented is a fact, even though the words "it could be" and "might be" are used repeatedly. On science program channels, a presenter's name often includes a title such as "theoretical cosmologist" or "theoretical physicist". The science channels sometimes use celebrities and even comedians to add credibility to scientific information that is being presented. Does this increase the certainty that the audience will be more receptive to the theoretical science *facts* being presented? Is this the intent of the producers?

Are producers influencing their viewers to accept theory as empirical truth, regardless of the educational consequences? Are some theoretical scientists and producers of video streaming trying to make people believe theory is fact? Is this a theoretical deception that "could" or "might" increase as world-wide video streaming continues to be rolled-out to the public?

Most viewers do not question something outside the realm of their understanding. It would seem they do not have the time or understanding to research the many theories being introduced by the world of science.

The same is true for high-end, make-believe graphic movies with embedded actors appearing in intergalactic plots. Some films introduce foreign gods claiming creation capabilities. How will this add to the mix of creator alternatives the new generations will have as they live their lives in the digital age?

Recently at a major university, two theoretical scientists were trying to solve a natural phenomenon that interested them. At the same time, they were trying to fulfill their need for a study that would qualify for a federal grant. The scientists observed houseflies with a full set of wings just before the flies' death, while also noticing other houseflies who died with partial or worn wings.   Before the flies' deaths, which were approximately four weeks, all of them still managed to fly.

The scientists applied and received a grant to set up a controlled experiment consisting of four progressive cages representing four weeks of a fly's life.   They primed the experiment with larva in the week-one cage. Each week they moved the houseflies along to the next week's cage.

The scientists tried photographing the flies in week four but found their equipment inadequate.   At this point, the scientists applied for a larger federal grant to obtain more expensive equipment for their experiment.   The grant was approved.   They purchased high-speed cameras that could photograph at 50,000 frames per second.   This made it possible for them to observe, in extreme slow motion, the minute detail of the houseflies' wings while they were flying.

What they found from their study was astonishing. The experiment showed that the houseflies with the worn-out wings adapted and could still fly by using only their legs for flight.   The flies moved their six legs back and forth much like a helicopter, but in alternating directions. This answered the question of why the flies with worn-out wings could fly, but it did not answer how.

After further study, the scientists discovered that the houseflies with worn wings could flex the exoskeleton top side of their legs rapidly while in flight, converting them into airfoil airplane wings providing lift. By rapidly reversing the shape of the top of a fly's wings it could remain airborne.

The scientists recognized that the experiment might have commercial and military applications. They immediately submitted a new grant asking for $1.2 million, which was approved. They procured additional equipment and staffing to cover the expected higher costs. The scientists also signed a contract with a television science channel for three-quarters of a million dollars for a three-part series starring Hollywood celebrities as commentators.

Believable? To some it may be, especially if presented as a fact – even if only a theory. However, in this case, the entire fly story is a fabrication, although it sounds highly believable by some people who are not biologists.

It makes the point that people *want to believe*. It also points-out that theoretical science is very important. People want to believe things that sound like they could be true.

Theories End Up in Textbooks as Facts

Theories sometimes end up in science books as facts. Science channels and classroom textbooks never include words declaring that the material presented is a theory. Should they?

Boards of Education across the US do not caution parents and students that certain material in their approved science textbooks may be theoretical. Should warning messages like the following be in science textbooks?

Some material presented in this textbook is theoretical.

Although it may be highly likely that someday it could be proven true, today it cannot be proven using empirical science.

It is important for students and parents to understand this because sometimes science findings may be drastically changed, and even reversed over time in both the theoretical and even empirical sciences.

Theoretical and sometimes even empirical science findings have been overturned

Theoretical Science is Needed

Theoretical science plays a necessary role in the experimental studies of life and nonlife mysteries found in the Universe. Most minds are naturally curious, but they are looking for truths. Theoretical science provides possibilities for the curious, but not necessarily facts and proof. Even so, there is a tremendous benefit in challenging people's minds to think outside-the-box.

In many cases, theories can never be observed. An example is the big bang. How would it be possible to go back 13.8 billion years and observe the big bang happening? It would not be possible. But there is no reason that the theory is not plausible. With the use of sophisticated mathematical data and related observations, there is consensus that the big bang theory is most likely true.

Theoretical science

A system of acquiring knowledge using mathematics, physics, statistics, or other science invented tools to explain something that is without factual proof

---

Some People Cannot Discern Theory from Fact

Decisions are no longer questioned by school boards when approving science text books which contain science theories for classrooms, presumably because of a lack of subject matter understanding. The boards most likely assume the theories are facts and rely totally on the guidance of scientists, governments, and sometimes legal courts.

The big bang is still a theory and should be taught as such even though it is widely accepted and most likely factual. The theory could be reversed someday or modified. Yet, the big bang theory is taught throughout the world's education systems as though it is a proven fact. People cannot discern the difference between theory and fact, especially when they have not been informed of the difference.

Because theories are always subject to change, should it be mandatory to explain to students and parents the difference between theories and facts?

Part II

Chapter 5

Did Aliens Create Humans

Almost half of the people in the world believe aliens exist. Some believe aliens created living things. Others believe they are just around to observe humans, and perhaps occasionally steal a dog or maybe a human on a dreary rainy night.

Very large outer space radio listening dishes are scattered around Earth and are trained towards the skies listening for noises coming from extraterrestrial aliens. Some enthusiasts believe aliens on distant planets may be talking to themselves or may be intentionally reaching out to other planets. They hope it is possible to someday overhear alien conversations with these electronic listening devices. These dishes have been operational for over 50 years with zero success so far.

In 1977, there was a possible noise signal coming from somewhere in outer space and picked up on the scientist's fine-tuned instruments. The noise was picked up from a sound dish at a major university and lasted for 72 seconds. The sound could never be deciphered successfully. Nothing has been heard like this from outer space since.

Meanwhile, there has been plenty of noise generated on planet Earth in the form of either analog or digital noise. Could it be, after hearing the turmoil on this planet, aliens have decided not to communicate with Earthly beings?

Do People Believe in Aliens?

There is a lot of hype about aliens in North America and, for that matter, throughout the world. People want to believe the images of aliens that have been portrayed in movies and on science channels. Fifteen percent of all books sold are about aliens and UFOs – all under the category of fiction. Not non-fiction. It seems people want to believe extra-terrestrial life really does exist.

If aliens do exist, they must be observing planet Earth at great distances. If aliens are close-by they must have technology greater than that on planet Earth enabling them to be unseen and travel at speeds unheard of today. The number of alien believers in the United States is slightly fewer than the rest of the world but still represents an amazingly large number. More people in the UK believe in aliens than God.

In the UK
52% believe in aliens, 44% believe in God

In the US
34% believe in aliens, 74% believe in God

Complicating matters there are examples of possible alien life on this planet which puzzle many. In Nazca, Peru, there are remnants of what people presumed to be a space station airstrip on the top of an arid mountain. The strips were carefully laid out in very long lines and included scorpion and bird art work. The people from this region have long believed extraterrestrial beings identified this area as a landing place for their spaceships. Many others around the world believe the same.

Since the discovery of the air strips by the rest of the world, the alien story has been refuted. It was determined these remarkable art forms were built a long time ago by very skilled local craftsmen using elevated perches to direct the work of laying rocks in intricate lines and patterns. Even so, many people continue to believe this was the work of aliens and find more evidence including a recently posted streaming video report of a strange "alien" at a gas station in Peru.

"Alien" Landing Strip in Peru

The United Nations Believes in Aliens

Although scientists and presidents have said aliens have not visited Earth and have not communicated from outer space in the last 50 years, the United Nations (UN) decided to include alien existence in their agenda. It seems the UN believes in aliens.

An astrophysicist was appointed recently to head-up the UN's new Office for Outer Space Affairs (UNOOSA). The appointment was made because of recent telescopic discoveries of seven planets in the habitable zones of other Solar Systems within the Milky Way galaxy.

Perhaps the UN thinks recent discoveries have made the appearance of extraterrestrial life very likely in the near future. UNOOSA wants to be Earth's first contact with aliens and coordinate humanity's response when aliens approach Earth. UNOOSA feels they understand the sensitivities involved and will be able to communicate with extraterrestrial life. Should humanity hope aliens land next to the UN building in New York City where UNOOSA resides rather than some rogue nation?

Legal Experts in Outer Space Law

There are legal experts in space law who have consulting contracts with the UN to advise them on the legal implications of space activities. They believe the first meeting with aliens will be by radio signals or microbes instead of meeting directly with aliens who may or may not be able to communicate with humans.

NASA Going to Mars

As part of NASA's upcoming inter-planetary mission to Mars, a new planetary protection officer position will be making advanced plans to protect the spreading of microorganisms between planets carried by human space craft and robot travel. The position requires recognized subject matter experience plus advanced knowledge of planetary protection.

Could Aliens Start Life on Earth?

If aliens created Earth's living humans, animals and plants, how did they do it? This type of creation could only happen after the Universe and planet Earth already existed and Earth was no longer volatile. This would have drastically narrowed the aliens' amount of time for creations compared to the creation timeline of the Universe.

No one seems to be making the argument that aliens created the big bang, Universe, stars, or planet Earth. Most alien talk is usually about aliens creating living things on Earth.

No one theorizes that aliens created the big bang or the Universe

. . . only living things

Why?

The most common theory today is that aliens did not actually visit Earth, but instead sent a meteor filled with amino acids that hit Earth. The meteors are alleged to have carried some form of life inside them. The life form was theorized as amino acids buried inside a meteor, traveling through space for billions of years in extreme cold and eventually colliding with barren, but stable Earth.

Were these meteors intentionally pointed and hurled at Earth with extreme precision with the intent of planting life on planet Earth? This would have been a challenge for the world's best aeronautical engineers. They would have needed the latest software to calculate the proper trajectory of a meteor to get it pointed in the right direction and at the right speed. This would have had an extremely high probability of failure.

> **Did aliens somehow send a meteor**
> **filled with amino acids**
> **that traveled through space for billions of**
> **years, without dying,**
> **to intentionally plant life**
> **at a predetermined target called Earth?**

Amino acids have been found inside meteors that have struck Earth. By themselves, it is impossible for amino acids to create living cells necessary for life.

Living Cells Require Information to Exist

Amino acids are a requirement of protein molecules for survival. Protein molecules are one of the major components of every living cell and are extremely complicated. This will be explained in Chapter 10.

> **Amino acids require
> information, laws, and systems
> to create living cells
> out of dead atoms**

For protein molecules to exist and to support life, they need more than amino acids. They need information and systems. Without them, living cells cannot exist.

Scientists Say Aliens Never Visited Earth

Aliens and UFOs have been studied by scientists for many years. These studies have shown there is no solid evidence that aliens ever visited Earth.

It might be difficult for scientists to accept aliens as creators because if they did, it would be like saying a pseudo-god was involved in creations instead of evolution or random events. So then, how could scientists consider a non-empirical creator agent such as an alien? Most scientists rule out aliens as a creator candidate for planet Earth, especially with no evidence as proof.

Are There Other Planets Like Earth?

Astronomers have been searching the skies for years to find life on other planets like Earth. Recently, they found a planet *in another Solar System* inside the Milky Way that possibly has life on it. They say the new planet is similar to Earth and sits in the habitable zone of its Solar System surrounding a star.

The problem with visiting this planet, which is 1400 light years away, is it would take 100 generations of people aboard an interplanetary rocket to make the journey. It would be like Noah's ark at sea for 1400 years with mandatory breeding to sustain life. Burials in space would be common. Hopefully, enough research would be conducted to determine if this planet is truly habitable before a rocket takes off with humans on board. Who would be willing to take this long journey?

The major differences between this newly discovered planet and planet Earth are that the new planet is about three times larger than Earth and much closer to its sun. The planet orbits every 37 days instead of 365 days and never rotates. One side is always bright at 160 degrees Fahrenheit while the other side is always dark with temperatures at 25 degrees Fahrenheit below zero.

Scientists estimate as many as twenty percent of the solar systems in the Milky Way contains planets that are in habitable zones of their own solar systems. They say with an estimated 200 billion solar systems, there may be as many as 40 billion planets that have the potential for life.

No one knows for certain if life beyond planet Earth exists. If life does exist, did aliens create the living things including humans or are aliens just a subject that some people enjoy pondering? Should research of aliens launching meteors carrying amino acids, or UFOs carrying aliens be scientifically funded?

Chapter 6

Is Evolution Creation

Evolution theory, as the creator agent for all living things, is generally taught as a *fact* in most schools. In reality, the theory of evolution has nothing to do with creations, or what is more commonly called the origins of life. Creations are the *beginning* of the evolution chain of events for each specie. While the true origin of life is being pondered by scientists, and theories are been proposed, the focus of evolution theory is now shifting to how life evolved *after* species were created.

Creation

The act of bringing something into existence out of nothing

Evolution is not the central theme of this book. Why? Because *evolution* has nothing to do with *creations*. This book is about creations. Since the 1960s, the term evolution has been used by the intelligentsia to educate students in public schools on how all living things were *created*. It has been backed by the US courts, who made their rulings after hearing arguments on only the *intelligent design* aspect of creation. Intelligent design was ruled as a part of religion. Their ruling stated that only empirical science findings could be taught in science classes of publicly funded schools. However, there was no prohibition from teaching creations by other creator agents such as aliens or God in civic or philosophy classes.

The US Supreme Court never ruled on the empirical aspects of evolution. Should they have?

As a result, evolution without empirical proof, is taught in all public educational institutions today. This is not to say that evolution should not be taught. It should be taught. It is only to say that evolution, as the sole creator of the Universe and all living things, should not be assumed and taught as if it were fact, and already proven, just because all other alternatives have been eliminated in the courts. This seems to be a very narrow view considering the minimal scientific evidence supporting evolution as a creator agent.

In Chapter 7 and thereafter, the term *random events* will be used in the discussion of the second creator agent proposed by scientists, instead of evolution, because evolution is a process that relies on *random events*.

Evolution 101 classes at prominent universities in the US are starting to teach that *theoretical evoluti*on has nothing to do with the creation of species. These classes now teach that evolution is the study of genetic changes that occur *during* the evolution of species*, not before.*

> **Evolution has nothing to do
> with the *creation* of a specie
> but rather
> genetic changes *after* a specie has been created**

Some may ask why it took so long for the intelligentsia to come to such an obvious conclusion, after knowing for 150 years what theoretical biologists Darwin and Wallace had already acknowledged.

Will it take another 150 years to educate school boards and the general public that evolution is not the same as creation?

Scientists have used the term evolution to depict how creation began. This is beginning to change. Evolution biology is about how life changed after its origin. It has nothing to do with the origin of a genetic chain of events. It has nothing to do with the origin of a specie or species.

Evolution classes are being redefined to cover a broader range of topics such as; some types of evolution involve randomness, while other types involve natural selection and adaptiveness. Other classes teach that evolution does not always result in living species getting better, such as their ability to survive. Evolution results in genetic changes between two *generations* of living things and can happen slowly or quickly over time. It can happen over one generation or many. Human behavior can influence evolution changes. Medical advances may alter human evolution. Natural selection cannot sense what the future needs might be of a specie in the evolution process.

Evolution has nothing to do with creating a living thing out of nothing. Evolution is different because it involves the study of living things that genetically evolve from one generation to the next. According to the definition of evolution, it has been repeatedly misused for a very long time.

Evolution

A *change* in, not the *creation* of, the inherited characteristics of a single specie across generations

---

Evolution theory is a highly respected science and does an outstanding job of looking at how species and microorganisms have evolved. It remains a theoretical science field, as do other science fields, because there are many missing gaps in specie genome lineage. However, this does not mean evolution theories are not important, well-supported, and broadly accepted. Will any of these theories ever be proven empirically?

Evolution Within a Specie is Real

Anyone who makes simple family observations of grandparents, parents, children, and grandchildren over five generations can easily see remarkable evolution. Newer generations seem to be generally stronger, taller, smarter, better-looking, and usually healthier. However, poorer countries do not fare as well.

Perhaps these changes take place because they live in better climates, have better food, better medicine, and better education than previous generations. Human families seem to be evolving within their specie unless their environment is static.

Evolution Is Controversial

Evolution, as a basis for explaining creation of living things, gained popularity starting 150 years ago by Darwin's research theories developed in the Galapagos Islands. Over time, the theory was accepted as fact. Religions bristled at the mention of evolution because they said it contradicted the Bible's, Torah's and Quran's explanations of the creations of living things.

Some scientists have used the term evolution, not only when speaking of plants and animals, but also when speaking about creation of the Universe and its parts. Across Europe, people have adopted the Darwinian theory of creation more than any other part of the world, especially when including humans in the evolutionary chain.

Humans Evolved from a Different Specie		
	Yes	No
Iceland, Denmark, Sweden, Norway, France, Japan, UK	80%	16%
Spain, Germany, Italy, Ireland	70%	24%
Poland, Greece, Bulgaria	50%	36%
US	40%	40%
Turkey	25%	50%

After Earth was created, it was void of living things for over four billion years. This is a difficult period in the study of creation steps for scientists in constructing the past because of Earth's extreme volatility. Like Earth, all other planets in the Solar System, seemingly failed to produce life in this time-period as well.

No New Species in 300 Million Years

Today some theoretical paleontologists believe there is enough evidence to prove, even with missing links, that species have evolved into other species along Earth's evolutionary journey. Other scientists look at the missing links with skepticism.

Some theoretical scientists have been asked to supply evidence of genome change or inherited chromosomes that promoted evolution from one specie to another specie. No scientist has been able to do this. This question has been asked:

Are there examples
of genetic mutations
or evolutionary processes
which changed
information in the genome?

The answer given to this question was: *creation* of most major species happened long ago and has not happened in the last 300 million years.

Today's species are all descendants from
ancestors 300 million years ago.

There is a popular misunderstanding
of what evolution is
by the common man
that says fish turned into reptiles, etc.

No new species have evolved
In the last 300 million years

There are some scientists who say there is a popular misunderstanding by the average person about how evolution works. Some scientists say if there were evolution steps from one specie to another, it took place too long ago to track. This is contrary to what is taught in educational institutions.

Is it possible the creation theory taught about evolution may not be true anymore?

Do fossils found today merely show a lineage trail within their own specie? Some scientists theorize they have fossils which show a possible link from one kind of specie to another, but they have no accredited scientific proof currently.

Most theoretical paleontologists point to the Grand Canyon as one of the best examples of archeological history of planet Earth. Estimates place the age of Earth somewhere around 4.5 billion years. The Grand Canyon represents the last two billion years of Earth's history. Fossilized plants and animals are found in layers of deposited sediment.

Somewhere near the top of the canyon at about 300 million years is where scientists have found an abundance of fossil records. They theorize this must be the beginning of creations of some of the major living species. At least 200,000 of the more than five million species known to have existed over the last 300 million years appear to have been created during this time. Many of the species identified by fossil records are extinct today.

The Grand Canyon

Before this period around 1500 million years ago the Protozoa, Earth's first one-celled animal, is said to have appeared. Hard shelled animals appeared 500 million years ago. Humans appeared much later.

DNA Maps Our Ancestors Back to 250,000 BC

More recently, science researchers have documented the genetic mitochondrial DNA of 100,000 species, reaching a conclusion that 90 percent of all animal species that are *alive today* got their beginning around 250,000 years ago, after an unknown cataclysmic event on Earth that apparently wiped out many species.

**90 percent of all animal species
that are *alive today*
got their beginning 250,000 years ago**

These same researchers say that instead of species continuously evolving into other species over millions of years of gradual evolution, *most early animal species began* at this distinct point in time on Earth and evolved within the specie from that point forward.

How Did Evolution Begin?

Scientists cannot show how the evolution chain began for living things. There must have been a beginning. Did aliens or evolution create living things? In other words, what primed the beginning of the creation of living things out of nothing?

The Carbon Dating Tool

Carbon dating is a method used to determine how long ago a dead plant or animal died. Scientists discovered that when cosmic rays collide with nitrogen in the atmosphere that carbon-14 is formed. Carbon-14 is continually absorbed by all living plants and animals while they are alive. When living plants or animals die, they stop absorbing carbon-14 and start losing it at a known rate. To determine the age of a dead plant or animal, a comparison is made between the level of carbon-14 in living things compared to the carbon-14 level found in the dead things being examined.

Carbon-14 findings can be controversial and hard to accept for young Earth creationists who believe Earth was formed somewhere between six and ten thousand years ago.

When an old book is found that was made from a living tree's fiber, the age of the book can be determined by the carbon-14 process. Fossil records are found continually throughout the world and carbon-dated to determine their age. A 44,000-year old cave in South Africa is revealing records of well-preserved artifacts used in the daily lives of its inhabitants.

Carbon dating

- Cosmic rays collide with nitrogen in the atmosphere forming Carbon-14
- C-14 is absorbed by all living things
- When living things die, they stop absorbing C-14 and begin shedding their C-14 levels at a known rate
- The current level of C-14 in something dead is compared to the C-14 level of living things to determine how long ago it died

---

In a cave outside Jerusalem, sets of homo sapiens teeth have been found and carbon-dated back to 400,000 BC. The teeth were in excellent shape and indicative of teeth found in humans today.

Evolution Isn't Powerful Enough to Create

Evolution theories proclaim that from a one-celled protozoa animal, all living things evolved over hundreds of millions of years from random mutations and natural selection. Some biologists disagree with this and say there would not be enough time in the 13 billion-year history of the Universe to create all the five million species that have been recorded. For certain it appears there would not have been enough time in the last 300 million years on planet Earth.

> **An early evolution theory claimed that
> from a single-celled Protozoa
> evolved all living things on Earth today
> over millions of years
> all by random selection.**

Evidence has been found by scientists that protein molecules, the basic requirement for living cells, are different for each specie. They found that it takes 139 protein molecules, plus information, to create a single-celled Protozoa animal. They say it also takes over 100,000 different protein molecules to create human beings.

- **Protein molecules are different for each specie. It takes 139 protein molecules plus *information and systems* to create the simplest Protozoa animal on Earth.**
- **There are over 100,000 different protein molecules needed to create humans, and there are over 5 million known species**
- **If it took one thousand evolution steps for a single protein molecule, it would take trillions of steps to evolve all protein molecules needed for all species**
- **There would not have been enough time to evolve humans in the 300 million available years out of the 4.5 billion years of Earth's existence**
- **Scientists cannot explain the evolutionary paths for either protein molecules or living cells**

With as many as five million species and 100,000 different protein molecules required for cells of the most complex species, the number of evolution steps required to create molecules for *all* species would be daunting. It would literally take trillions and trillions of steps with absolutely no chance of a single failure between steps to evolve protein molecules.

If all these protein molecule creation steps were combined, it would have taken far more time than the 13.8 billion-year age that the Universe has been in existence.

After 60 years of teaching evolution as a creator agent, sixty percent of all public-school teachers still struggle to explain how the creation of atoms came about *out of nothing*. Mandatory retraining courses are being designed to help them wrap evolution around the evolution creator agent explanation. At the end of this re-training will these courses be enough for the teachers to explain creation out of nothing?

After 60 years of public-school teaching that evolution is a creator agent

60% of teachers still struggle to explain how creation of atoms *out of nothing* came about

**Newly developed training courses intend to show them
*"What's Causing Creations"***

Will these new courses work?

Should theologians take creation science courses to complement their teachings from the Bible? Would this help bridge the informational gap for young students, as well as adults, explaining how faith and scientific creation complement each other? Evidence of theologians acting on this seems difficult to find. Is congregational backlash the problem?

Should theologians consider taking a creation science course?

Scientists cannot explain how a single protein molecule evolved. They cannot explain the evolution path of even the simplest of all one-celled animals. Creating multi-celled living animals and plants is far more complicated than anyone can imagine.

Will anyone ever figure out how to prime the beginning of the evolutionary chain for each specie?   Or said another way, will anyone ever discover how to create living things out of nothing but dead atoms?

Chapter 7

What Created the Big Bang

If an alien is not a creator agent, then only two options are left: random events (evolution) or God. Again, evolution makes use of random events and is not a creator agent (see Chapter 6).

Creator Possibilities

Aliens
Random Events
God

This begs the question, what is a random event and how was it created? Also, how does a random event know how to create? Perhaps these questions sound odd, but they are important. If random events are powerful

96

enough to be a creator, then they need to be understood more.

Random Event Creation is Considered Scientific

It is evident that *random event creation is considered scientific* because it is financially supported by private and public institutions and is not challenged by the courts. Why is random event creation considered scientific? Looking at the definition alone, random event creations is not very reliable.

**Random event creation
is considered scientific today
even though it is unreliable**

What is a Random Event?

A random event is any event that *could* occur or may *never* occur. It cannot be predicted. For certain, if a random event did occur, as a creator agent, it was not influenced by anyone or anything.

Random Event

An event that may or may never occur,
cannot be predicted,
and is not influenced by someone or something

What is the Big Bang?

Theoretical scientists say the big bang occurred 13.8 billion years ago, creating the Universe and billions of galaxies, stars, and planets. They have calculated the timetable of specific events that must have happened during the big bang, including some of the events that happened *before* the big bang began.

Their big bang timetable shows random event creations of pieces of atoms occurring in steps as little as a trillionth of a second. The big bang did not happen as just one step. Multiple random events would have been involved. In fact, trillions of random events must have been involved. A single random event would have to have extreme discipline to successfully create even one of its deliverable steps during the big bang.

One of the characteristics of random events is *they may never happen*. If just one step of the big bang timetable was interrupted, that missed creation step would have the potential of stopping completion of the entire big bang and the Universe. Every creation step that followed, including the creation of Earth and all living things, would have never happened.

Pretend *YOU* Are the Random Event

If random events are responsible for creations, then to understand this more fully, it would be advantageous to look at a random event through a human perspective. So, pretend you are a random event. In this way, you, the random event would reveal all the creation steps it used to design, create, and operate the Universe.

Creations of the Universe, Earth and living things by random events, it seems, would have to incorporate humanistic thought processes like decision making. This could not be accomplished by a simple machine or robot, even with artificial intelligence. Artificial intelligence is a product of mankind, so it is ruled out as a tool that a random event could use. Eliminating thought processes, like those that humans employ while making decisions, would make random events, by definition, not a creator agent.

The Big Bang Timetable

1 million years later	First stars began to shine **Hydrogen** and Helium atoms created
1 hour later	**Helium** neutron created

0 time   --- *The Big Bang* ---

1 millisecond before	**Hydrogen proton created**
1 picosecond Before	**Creation of 4 fundamental forces in matter created (gravity, strong force, weak force, and electromagnetic) plus quarks, bosons, science laws, systems and information**

In the very beginning of creations, before the big bang, random events would have started out in a dark, deep space completely void of everything. There would have been some sort of real estate involved. Call it outer space without monument markers. This is where the first step of *creating something out of nothing* begins. Scientists have assumed that dark matter, gravity, energy, boson matter, science laws, systems and information already existed in empty space, somehow, even before the big bang. How did random events create all the precursors to the big bang?

**If random events could not
use thought processes *in creation*
(like those used by humans)
it would make them non-functional**

Are theoretical scientists continuing their work without taking into consideration all these precursor creations that must have occurred before the big bang? Have they given up explaining these things because they cannot explain them?

**How did random events
create everything that occurred
prior to the big bang?**

A random event would need to decide ahead of time where to place this newly-created Universe in outer space. Random events would have to be intentional;

otherwise, how could a planet like Earth and its inhabitants exist? But if random events are intentional, then they cannot be a random event.

If random events are intentional they cannot be a random event

---

A random event would need to know in advance, or acquire a skill set to know, how to design an atom as well as the systems and laws of science that support it. The random event would also have to have knowledge about how to design and operate a perfectly designed perpetual atom machine.

An atom is by far the greatest example of a perfect design. Atoms never stop. They slow down a bit and speed up a bit due to cold or heat, but they just keep on going . . . no batteries . . . no winding up . . . no human intervention . . . no maintenance . . . a perfect machine.

Before the very first atom was created a random event had to have designed and created systems and science laws for an atom to be operational. These creations would then become the information and operational glue holding this newly created atomic invention together, including the perpetual operation of it.

When expanding the design of the first atom into much more complicated atoms, advanced designs would be necessary to create them. This would be especially true as multiple atoms were joined together forming molecules. Each atom and molecule would have to be designed ahead of time with specific planned uses, functionality, weights, tastes, and colors, just to name a few.

Everything in the Universe is entirely made from atoms. That includes humans. Every hair. Every bone. Everything.

A random event would have to think ahead to design a gold atom to make certain it could be used for such things as decoration or electrical conductivity. Carbon and oxygen atoms had to be designed ahead as food for living things.

About now it is obvious that it would take a very special random event to design atoms, and this is just the beginning of the design events required to create a complete universe.

Why would a random event design and create an oxygen atom with no intended purpose?

Designing Without Advanced Planning

Designing and creating without advanced plans for each new creation step that pops-up during the creation of the Universe does not work. There must be continuity of all the steps from the creations of the big bang to the creations of living things. The Universe is the perfect result of a continuum of uninterrupted creation steps without failure. All new creation steps were serial. Some could have been parallel, but the important ones were serial.

There is no chance for error in serial continuum creations, otherwise there is failure. The creation steps number into the trillions. There is no chance for error or the chain of events are broken, and the remaining creation steps will never happen. This is especially true at the time of the big bang when the foundations of the Universe were built. If random events were responsible for the distribution of atoms, then why don't all planets have the same kinds and amounts of atoms? Was there selective distribution? If there was selective distribution of atoms only for Earth, then how did random events accomplish this and why? (see Chapter 9)

All creation events are measured by the probability, or odds, that they might occur. When talking about big numbers, scientists use what is called scientific notation.

Probability

A branch of mathematics dealing with the *chance* that an event will occur

---

If scientists talk about the probability, or chance, that something may occur as being very remote, they might say one chance in 10 to the [10th] power chances, or 1 in 10^{10th}.

This is shorthand for one chance in 10,000,000,000 chances. This is a big number that needs to be abbreviated; otherwise, pages would be filled with zeros for events that have remote probabilities of ever happening.

Here are some examples of scientific notation:

10^3 = 1,000

10^9 = 1,000,000,000

10^{81} = 1,000,000,000,000,000,000,000,
000,000,000,000,000,000,000,
000,000,000,000,000,000,000,
000,000,000,000,000,000

To show the impact of probabilities in our daily lives, during the 2014 NCAA men's basketball tournament, a successful business philanthropist made a one billion dollar offer to anyone in the world who could fill out the NCAA basketball bracket with 100 percent correct winners and scores.

Many people tried but no one won. The person making the offer knew the odds of getting all the correct answers was one chance in 10^{18} chances, or 1 chance in 1,000,000,000,000, 000,000 chances.

Seven billion people on planet Earth armed with super computers could not come up with the bracket answers filled out correctly. The person who made the offer was no fool because it was impossible to lose the money knowing the odds of winning.

Greater Than 10^{50} Will Never Happen

Long ago a mathematician theorized that a single *complex* random event greater than 1 chance in 10^{50} chances will never happen. Most of the individual steps in the creations of the Universe exceed these odds. Steps that have less than these odds usually are a part of a series of serial event that when taken altogether result in odds much higher.

All creation events are *complex random events.* A *simple random event* would be like the lottery. The 33-state lottery is difficult to win, with odds at one chance in 10^8 chances, or one chance in 195,000,000 chances of winning. If a person bought one-hundred million one-dollar lottery tickets they would still have only a 50 percent chance of winning! Today state lottery commissions manipulate the chance of winning, jackpot size and increased participation by simply adjusting algorithms.

Some scientists argue that standard statistics used to measure complex creation probabilities are not valid. They say a modified version of statistics developed for scientists to do theoretical quantum physics calculations gives more correct views of the Universe and negates standard statistics. However, standard statistics is the backbone tool used to support all medical research today.

Philosophers argue that even with a 10^{50} impossibility rule there is still that *one chance* that a random event could still happen.

How big a number would it take for some people to believe before a random event will never happen? Is it one chance in **10^{100}** chances or more?   How large a number does it take before a creation event could never happen randomly?   This book shows the very large numbers that it would take if random event creation was successful.

Remember, each of the creation events or steps are *complex* serial events with each event relying on the completion of the preceding event, or the next event will never occur. There are literally trillions of complex creation steps making up the creations of just the big bang. None of these creation steps ever failed.

How Big is 10^{50}?

Can you imagine a roll of one-inch long, 50-50 raffle tickets wrapped around the world 700 billion, billion, billion, billion times? With only one try, could a winning ticket be selected? Could anyone win? That's the kind of odds of winning the mathematician clearly understood - one chance out of 10^{50} chances or greater could never happen.

From the creation of an atom to the creation of humans, some scientists theorize that random events are the only possible creator of everything that ever was or will be. Are they correct?

Can't Happen Rule

Complex random events with a probability greater than 1 chance in 10^{50} chances will never happen.

10^{50} = one chance in
100,000,000,000,000,000,000,000,000,
000,000,000,000,000,000,000 chances

If each chance was one-inch wide
it would wrap around the world
700,000,000,000,000,000,000,000,
000,000,000,000,000,000 times.

What It Took to Form the Universe

There were basic precursors that had to have been met before the big bang could begin. How else could something so complex happen with such cosmic results?

Creating and operating a universe has similarities to creating and operating a computer. Both are extremely complicated and require precision. Both rely on previous steps successfully created and operated, or the subsequent creation and operational steps cannot occur.

Here are a few of the basic design parameters that a random event creator agent needed to comprehend in creating the Universe.

- **All creation events were complex events**
- **All creation events were executed serially**
- **The laws of science and systems were created prior to the first creation event**
- **Complex random creation events with a probability greater than 10^{50} should never have happened, but they did**
- **None of the trillions of serial creation events ever failed**

Each of the millions of components and gates inside a computer must perform perfectly over a life span of six years, with good luck. A universe must do the same for at least 13 billion years.

Digital data travels through circuits mostly in serial paths. To increase reliability, computer manufacturers sometimes build parallel paths, which increases reliability but also increases costs.

Predicting success rates for the operation of a computer involves multiplication of each of the individual component's success rates times the success rates of all other components.   If the total rate is less than 100 percent perfect the computer will not run very long. This is also true for the operation of the Universe.

Creating and operating a Universe
is much harder than creating and operating
a computer over a span of 6 years

The failure rate of a Universe
must be zero over billions of years
in order for life to exist today

If any of the creation or operational steps had failed along the path of the Universe's creation, there would be no Universe today.

Obviously, the Universe's creation steps had extremely high success rates, or it would not have lasted over 13 billion years.

Creating a Universe is Very Complicated

Whatever created the Universe had to be very powerful and highly intelligent. Whether it was created by aliens, random events, or God, it involved powers which may never be empirically proven or intellectually fathomed by humans today.

Creating something out of nothing in an empty space baffles the brightest minds. Scientists studying micro-sections of the big bang creation make discoveries almost daily. None of the discoveries begin to come close to explaining *creations out of nothing* of the Universe and its living inhabitants.

Did a continuum of trillions of serial random events create something as complex as a Universe and living things without any failures? Does the definition of a random event make this impossible? A random event is not predictable. That means it might happen one second from now, one year from now, a billion years from now or never.

**Random events would not work well
for creating the big bang
as trillions of random events steps
were required to create the Universe.**

**By definition, random events sometime never
occur or occur so infrequently as to not be reliable
as the main creator agent for such a large project
as building a Universe.**

**Every one of the sequential steps of creating the
big bang and ending in creation of living things
had to occur inside the
scientist's calculated Universe timeline
with *zero* failures.**

The chance that a random event happens with design precision seems impossible because a random event is not influenced by the ideas or actions of anything. Because no one can predict when random events will happen, there is zero possibility that anything can influence the outcome without changing the meaning of random.

Creations Should Never Have Happened

The laws of science and systems had to be created first including gravity, mass, energy, and force. They all had to be in place before the big bang could be created. Some theoretical physicists and cosmologists have calculated the random event chance of these precursors being in place before the big bang could begin as the following:

	Chance
Laws of science and systems	10^7
Strong nuclear/electromagnetic force ratio	10^{16}
Ratio of gravity and electromagnetic force	10^{40}
Mass density value	10^{59}
Space energy density constant value	10^{120}
Expansion rate of the Universe	10^{55}
Ratio of electrons to protons	10^{37}
Chance the big bang could have happened randomly	$10^{9,221,000,000}$

Is it possible, with the astronomical odds of the big bang happening randomly, and with all the pieces that had to be in place in the proper sequence, that the big bang events could have occurred randomly?

Chapter 8

What Created Atoms

Some scientists theorize that atoms were created by random events during the big bang. They say millions of years later atoms were randomly distributed throughout the Universe forming stars and planets. Many theories have been proposed about how creation steps were executed, starting with the big bang and ending with the creations of galaxies, stars, and planets.

Human beings are made from one hundred percent atoms. No fillers. How atoms were created is rarely talked about, and yet they are vitally important to the human body and the entire Universe.

For example, a single human cell contains 10^{80} atoms. This is an extremely large number of atoms inside just one cell. Knowing there are 13 trillion cells in an adult body with 10^{80} atoms in every cell, gives an indication of how difficult a job it would be to calculate the total number of atoms in the Universe. Some scientists have tried to calculate this number, but their answers are not considered important enough for debate.

Creations Are Complicated

Every atom in the Universe is identical to every other atom of its type, no matter where it is found. A hydrogen atom found on a faraway planet is identical to a hydrogen atom found on Earth. Atoms are all made from the same interchangeable pieces called electrons, neutrons, and protons.

Atoms rely on invisible things called science laws, systems, and information to exist. These allow atoms to perform endlessly. This is rarely discussed by scientists or theologians when they talk about creations.

Besides the atoms found in our bodies, atoms are everywhere. Everything you see, touch, taste, hear, or smell is made of atoms. Everything that is dead or alive is made of atoms. It does not matter whether it is a rock, sunflower, horse, human, or water. They are all made of atoms.

Atoms on planet Earth are plentiful and are found in the exact types and quantities necessary to support life.

Less than one percent of space in the Universe is comprised of atoms. The remainder is nothing but empty space. The vastness of space seems unimaginable when comparing the relatively small size of planet Earth, or even the Milky Way galaxy to the immensity of the Universe.

The origin of atoms and their role in the Universe is a wonderment subject introduced into seventh-grade science classes. The subject is covered in about two sentences with offering only the big bang evolution explanation as to how atoms came into existence.

Atoms Are the Building Blocks of Everything

By rearranging the electrons, protons, and neutrons of existing atoms, new atoms can be built by scientists. The atomic table contains the names of about twenty new atoms scientists have *produced* by rearranging atom pieces. This is *not creation*. This is *production* of new atoms, but still an amazing feat that only scientists can accomplish.

Atoms can become other atoms or particles as they decay. These particles never disappear. They remain forever and are recycled.

A molecule is comprised of two or more atoms. A cell is made up of millions of molecules. Infants begin with a few cells and create millions of new cells daily during their growth period. As adults naturally age and approach death, they lose millions of cells daily.

What Are Atoms?

Atom is a word used 2500 years ago by a Greek philosopher and mathematician called Democritus. He coined the term after picking up a handful of sand and removing all the grains except one. He crushed the remaining grain of sand between two smooth surfaces and then eliminated all the sub-grain particles except one.

Anuokpitoc Democritus

118

At this point, Democritus could barely see the remaining particle. He theorized if he continued the process, the particles would get smaller and smaller until only the smallest invisible particle remained. He called his invisible theoretical particle an atom. It took 2300 years for his theory to gain interest with other scientists.

Sometime during the big bang explosion, it is theorized that a hydrogen proton was created followed by an electron, allowing for the creation of the first complete hydrogen atom. Hydrogen is the only atom without a neutron in its nucleus.

After the hydrogen atom was created, a helium atom was created. This atom had a new particle added to its nucleus called a neutron giving it two electrons, two protons, and two neutrons.

According to the big bang theory, the first atom particle created was a Hydrogen proton and was followed by the creation of a Helium neutron. On the surface it looked like a simple creation, but in fact, it was very complicated.

The Big Bang Timetable

1 million years
later

First stars began to shine
Hydrogen and Helium atoms created

1 hour later

Helium neutron created

0 time   --- *The Big Bang* ---

1 millisecond
before

Hydrogen proton created

1 picosecond
Before

Creation of 4 fundamental forces in matter created (gravity, strong force, weak force, and electromagnetic)   plus quarks, bosons, science laws, systems and information

---

The big bang theoretical timetable does not mention at what point electrons were created. The assumption is they were previously created and available on demand to complete the creation of the Hydrogen and Helium atoms as well as other atoms that would follow.

Also, it should be noted that in the big bang chart, there are events that began *before time zero* such as the creation of science laws, systems, and boson sub-atomic particles to name a few.

Some theoretical physicists have stated there was no God creator during the big bang because time stood still and therefore God could not exist. Based on the events in the timetable it appears this is not true.

It is apparent from just looking at the theoretical timetable presented by scientists that there were many events prior to the big bang, which means time did not stand still as theorized.

Ninety-nine percent of an atom is empty space. The neutrons and protons are in the middle forming the nucleus. The nucleus contains ninety-eight percent of an atom's weight. The other two percent of the weight is in electrons that circle the nucleus.

Electrons spin around the nucleus at 186,000 miles per second. Amazingly, at this speed the atom always stays together and does not separate or collapse. Science laws and systems make it possible for this creation to happen.

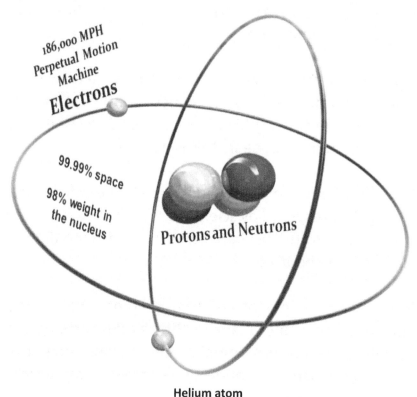

Helium atom

As the *blueprint* for creation of an atom was completed and the atom was created, hundreds of different atoms were then created and distributed throughout the Universe. These atoms have been identified and listed in today's atomic table. The first eight atoms are lighter than air. In fact, they float in the atmosphere. Humans breathe atoms into their lungs continually.

The mass, or weight, of each atom is shown in the atomic table.

No	Name	Sym	No	Name	Sym	No	Name	Sym	No	Name	Sym	No	Name	Sym
1	hydrogen	H	25	manganese	Mn	49	indium	In	73	tantalum	Ta	97	berkelium	Bk
2	helium	He	26	iron	Fe	50	tin	Sn	74	tungsten	W	98	californium	Cf
3	lithium	Li	27	cobalt	Co	51	antimony	Sb	75	rhenium	Re	99	einsteinium	Es
4	beryllium	Be	28	nickel	Ni	52	tellurium	Te	76	osmium	Os	100	fermium	Fm
5	boron	B	29	copper	Cu	53	iodine	I	77	iridium	Ir	101	mendelevium	Md
6	carbon	C	30	zinc	Zn	54	xenon	Xe	78	platinum	Pt	102	nobelium	No
7	nitrogen	N	31	gallium	Ga	55	caesium	Cs	79	gold	Au	103	lawrencium	Lr
8	oxygen	O	32	germanium	Ge	56	barium	Ba	80	mercury	Hg	104	rutherfordium	Rf
9	fluorine	F	33	arsenic	As	57	lanthanum	La	81	thallium	Tl	105	dubnium	Db
10	neon	Ne	34	selenium	Se	58	cerium	Ce	82	lead	Pb	106	seaborgium	Sg
11	sodium	Na	35	bromine	Br	59	praseodymium	Pr	83	bismuth	Bi	107	bohrium	Bh
12	magnesium	Mg	36	krypton	Kr	60	neodymium	Nd	84	polonium	Po	108	hassium	Hs
13	aluminium	Al	37	rubidium	Rb	61	promethium	Pm	85	astatine	At	109	meitnerium	Mt
14	silicon	Si	38	strontium	Sr	62	samarium	Sm	86	radon	Rn	110	darmstadtium	Ds
15	phosphorus	P	39	yttrium	Y	63	europium	Eu	87	francium	Fr	111	roentgenium	Rg
16	sulfur	S	40	zirconium	Zr	64	gadolinium	Gd	88	radium	Ra	112	copernicium	Cn
17	chlorine	Cl	41	niobium	Nb	65	terbium	Tb	89	actinium	Ac	113	ununtrium	Uut
18	argon	Ar	42	molybdenum	Mo	66	dysprosium	Dy	90	thorium	Th	114	ununquadium	Uuq
19	potassium	K	43	technetium	Tc	67	holmium	Ho	91	protactinium	Pa	115	ununpentium	Uup
20	calcium	Ca	44	ruthenium	Ru	68	erbium	Er	92	uranium	U	116	ununhexium	Uuh
21	scandium	Sc	45	rhodium	Rh	69	thulium	Tm	93	neptunium	Np	117	ununseptium	Uus
22	titanium	Ti	46	palladium	Pd	70	ytterbium	Yb	94	plutonium	Pu	118	ununoctium	Uuo
23	vanadium	V	47	silver	Ag	71	lutetium	Lu	95	americium	Am			
24	chromium	Cr	48	cadmium	Cd	72	hafnium	Hf	96	curium	Cm			

100 Different Atoms Found on Earth

In addition to the differences in the weight of an atom, atoms take on different characteristics. Some are salty, some are sweet, and some are invisible. Others are heavy, shiny, or lethal.

What Designed Atoms?

Did random events somehow design and create atoms as individual perpetual motion machines that ultimately became the basic building blocks for everything? Did random events know ahead of time an atom's purpose? Was the atom designed? Were the science laws and systems that drive the atom designed and created by random events? Were they designed by aliens?

How then were atoms created? The answer theorized by scientists today is that during the big bang, random events made use of a subatomic particle, called the boson that had already been created. They theorize the boson gave *creation to atoms* by adding mass, or weight, to the atom's proton and neutron.

The Boson Particle

The Hadron Collider Project is one of the most intriguing science experiments today with 11,000 scientists from 100 countries participating. One of its main objectives is to discover if a theoretical subatomic particle called the boson, could be responsible for adding mass, or weight, to the proton and neutron nucleus of an atom. Other goals of this research include such things as examining the possibility of splitting off other subatomic particles for potential use in the treatment of diseases such as cancer.

Knowledge of atomic structure is complicated. Because atoms are invisible, it becomes a very specialized field of science requiring highly trained scientists and specialized tools.

Atoms
Are the Basic Building Blocks
of Everything

creations = dark matter bosons

bosons = electrons, quarks

quarks = protons, neutrons

electrons, neutrons, protons = atoms

multiple atoms = molecules

millions of molecules = cells

trillions of cells = humans

Scientists have theorized that at the time of the big bang, electrons, protons, and neutrons were created, but only as shell bodies without weight, or meat on their "bones". Think of an empty shell body being created during the big bang, with nothing on the inside, and later bosons being added like filling to a cherry pie. They theorize the boson subatomic particles were created sometime before the big bang in sufficient quantities to add weight on demand to protons and neutrons as they were being created.

The boson has sometimes been called the god particle because it is theorized it adds weight to an atom. Some would consider this the major part of creation, or even the creator.

The gluon, a strong force particle, is extremely useful magnetically because it holds neutrons and protons together. Another theoretical subatomic particle which adds no weight is the graviton. The graviton helps to define the science laws of creations.

Particles manifest themselves in ways other than atoms. Some are beams made up of charged particles such as photons, protons, electrons, pions, kaons, and muons. A laser beam is made of photons, allowing light to travel through space.

Subatomic Particles

Bosons = energy force particles (quarks)
   that give mass to protons and neutrons
Gluons = strong force particles using
   photons to magnetically hold together
   protons and neutrons
Gravitons = laws of physics of creation
   (gravity)
Photons = electromagnetic light energy
   carriers

The **Hadron Atomic Collider**

Scientists are conducting experiments inside a cylindrical magnetic tunnel that is seventeen miles long and buried 100 meters below the surface. It is straddling the Swiss and French border and is called the Hadron Collider. It's the world's largest atomic collider. This benefits science worldwide as scientists perform experiments from its output.

The magnets lining the walls of the tube require seven trillion volts of electricity to force the particles to the center. When the collider is turned on, lights dim outside in near-by villages. When the moon passes overhead, its gravity causes scientists to make necessary adjustments to keep the collider beams focused in the center of the tube.

Hadron Collider

Protons freed from hydrogen atoms are released in opposite directions and propelled in the center of a tube by magnets, whirling them around near the speed of light. When protons collide, subatomic particles are released for scientists to examine.

Because hydrogen is the only atom that contains only one neutron and only one electron, the proton is easily separated from the atom by a low-tech battery. This provides a simple way to harvest protons as input for the Hadron Collider experiment.

If the boson subatomic particle already existed before the big bang, then adding the boson to a proton or neutron shell would not be *true creation* and certainly not a random event. It would be *producing* an atom from parts that *already existed*. What caused the creation of the bosons? What caused the creation of the shells?

Scientists have not theorized how the boson subatomic particles were created. They also have not theorized where the neutron and proton shells came from or, for that matter, how atoms were assembled.

> **Adding the boson particle**
> **to a neutron or proton shell**
> **is not creation because**
> **the boson particle was previously created**

Science laws and systems were the final ingredient required to create fully functional atoms. No one knows from where they came.

Was an invisible thing, called a random event, powerful enough and smart enough to create an atom? Did aliens create atoms?

Chapter 9

Why Is Earth So Different

Over one hundred different kinds of atoms were created for planet Earth, but not created for other planets. In the Solar System galaxy, planet Mars received fewer types of atoms. In fact, every one of the planets in Earth's Solar System received a different assortment and quantity of atoms than did Earth. Yet these planets all came from the same big bang, the same Universe, the same galaxy and even the same Solar System. How can it be that Earth was created differently than the other planets of the same Solar System, if the Solar System was randomly created?

Other planets in Earth's Solar System are not as fortunate as planet Earth. A major staple they all lack is water. It is theorized by some scientists that a few planets may have had water but probably not in the quantity required to support plant and animal life then or now.

Compared to Earth, why does Mars have only a sub-set of Earth's atoms? Something created these atoms. Perhaps random events skipped over Mars too quickly when creating atoms?

If Venus and Mars are close to Earth and in the same habitable zone that promotes life, why aren't they inhabited by some form of living things like Earth? They are close enough to be observable with today's technology. Why are these planets mostly barren of water while Earth has plenty?

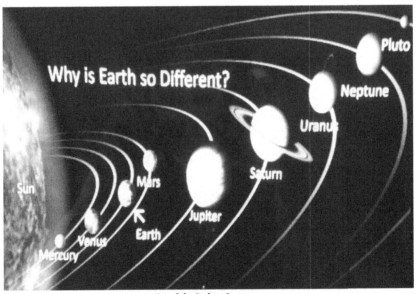

Earth's Solar System

If chocolate chip cookie dough is mixed in a bowl (the Sun) and globs of dough are distributed on a baking tin (the Solar System), would not all the cookies be chocolate chips? Why does Earth's Solar System have such a variety of planets such as chocolate chip, ginger snap, and peanut butter? They all came from the same Sun bowl, right?

Scientist want to explore planet Mars more closely to determine if life may have existed there. Currently, telescopes reveal there are no inhabitants on any of the planets in the Solar System except Earth.

Mars Landscape

Earth Looks Out of Place

Scientists speculate that it is possible some planets in Earth's Solar System may have been inhabited at one time. They say within a few million years after the big bang explosion, there were an unbelievable number of stars created. Shortly thereafter, an even greater number of planets were created. With all these events going on, is it possible that an isolated creation event like Earth happened *randomly* on another planet? Earth seems to look out of place with all its living things in comparison to other planets.

Earth Landscape Yellowstone

Many billions of galaxies were formed after the big bang, including planet Earth's Milky Way. Some scientists believe there may be other planets like Earth that were created but are thousands of light years away from Earth.

It is theorized the big bang set off an explosion shooting billions of stars outward from its center in all directions. The stars somehow were retained in a revolving flat platter called the Universe. The explosion formed galaxies, galaxies formed stars, and stars formed planets and moons.

With all the creations taking place at the time of the big bang, why was planet Earth's creation so different?

The Giant Cosmic Peashooter Theory

Did a theoretical *giant cosmic peashooter* help form planet Earth?

At the time of the big bang, immense piles of protons, neutrons, and electrons were theoretically being created out of thin air and formed into atoms. How many? Apparently, immense numbers that no one can calculate.

Is it possible that a giant cosmic peashooter, loaded with one hundred different kinds of atoms in the right quantity for each atom, shot the atoms through space for millions of light years and landed them in a habitable zone of one of the billions of Solar Systems scattered throughout the Universe? Of course, this would have happened millions of years after the start of the big bang, only when the necessary atoms for Earth had been created in abundance.

This theoretical *random event* peashooter would have been highly intelligent with design and creation capabilities.

 At a time when atoms were being created at a rate of trillions of atoms per picosecond, the peashooter, after gathering what it needed for Earth's survival, found the precise spot in the Universe. It then, theoretically, discharged the entire load with cosmic force.

Then imagine a gigantic catcher's mitt orbiting at the right speed and distance from its sun catching all the atoms and allowing them to take up residence in their final resting place in a ball called Earth.

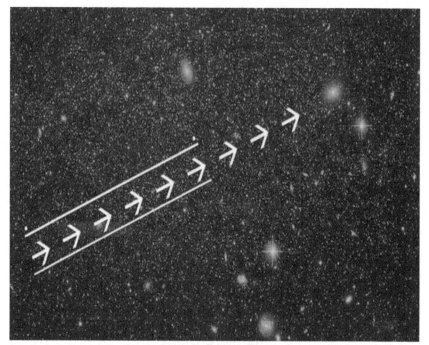

Cosmic Peashooter

Looking at Earth's unique location in the Universe, is it too farfetched to theorize that a random event pre-selected a premier location to plant such a desirable life-sustaining creation payload? Can anyone know for certain?

However, one thing is certain; planet Earth is drastically different from other known planets. There is speculation about life on other planets, but with the limits of space, time, and present technology no one can empirically prove life exists outside of planet Earth. The timetable for the creations of the Universe indicates that many major events happened, beginning with the theoretical big bang and ending with the creation of human beings. For simplicity, trillions of creation steps are left out of the following chart that spans an estimated 13.8 billion years.

13.8 billion years ago

\

Systems and science laws

Big bang

Universe

Earth

Protein molecules

Single cell animals

Humans

\

today

Science Laws and Systems

Systems, information, and science laws are essential for creations but rarely mentioned. One thing is certain. No one on Earth created them. Scientists discovered them and are documenting their use in physical science today.

But where did systems, information, and science laws come from?

Gravity is a good example of a science law. No one can change gravity's physical properties or the constant value it represents.

At the time of the theoretical big bang, gravity had to have been already created or there never would have been a big bang.

Systems

**A set of connected parts that use
information, science laws
and reliance on each other
to support a new complicated whole**

Systems are different than science laws because they contain logic and information. Scientists did not create them. They merely discovered them.

Science Law

**Theoretical statements inferred from facts
applied to a phenomenon that always occurs
if certain conditions are met
culminating in a science law**

The Anthropic Principle

In the 1970s there was a major worldwide collaborative study by scientists to understand how Earth came into existence as a habitable place for plants and animals.

It was theorized that Earth's living things must have had specific constants and values for living things to survive. Scientists believed the constants and values had to always remain true, without change.

The habitable zone, along with its constants and values, set the stage for a new term called the *Anthropic Principle*. It meant the exact ingredients, with proper quantities, had to come together at the exact time they were needed to support life, as it was being created on planet Earth.

Anthropic Principle

**When all the necessary ingredients
come together at a planet
in the exact amounts and at the exact time
life can be created from dead atoms**

After the Anthropic Principle was proposed, scientists from all over the world gathered to debate this new theoretical principle. The main question being asked, was how could such a beautiful green and blue planet, the only one in existence in its Solar System be so different than the other planets?

One by one, scientists identified over 200 constants and their specific values that were needed for life to exist. As an example, they stated that gravity must exist and have exact values to keep objects from flying away from Earth into outer space. If gravity were too strong, Earth would crush living things against its floor.

Some of the constants that scientists studied included water vapor levels in the atmosphere, atmospheric pressure, distance from the parent star, inclination of

Earth's axis, ozone levels, carbon dioxide levels, and thickness of the Earth's crust. All were vital to promote life on Earth.

Over time, the list of 200 was whittled down to 162. The constants and values have been agreed on and are readily known throughout the world today.

If any of the constants did not exist or their specific values were not exactly what was needed, Earth would be just another planet not fit for habitable living.

Atmospheric pressure	Poleward heat transport in planet's atmosphere
Average rainfall	Quantity and extent of forest and grass fires
Carbon dioxide level	Quantity of geobacteraceae
Distance from parent star	Quantity of iodocarbon-emitting marine organisms
Gravity interaction with moon	Quanity of phytoplankton
Hypothermal vapors in the upper crust	Quanity of soil sulfur
Inclination of orbit	Ratio of dual water molecules to single
Iron quantities in soil and oceans	Molecules in the Troposphere
Magnetic field	Soil mineralization
Nitrogen quantity in atmosphere	Solar wind
Oceans to continents ratio	Stratospheric ozone quantity
Oxygen quantity in atmosphere	Surface gravity escape velocity
Oxygen to nitrogen ratio in atmosphere	Temperature
Ozone level in atmosphere	Thickness of crust
Parent star mass	UV radiation at the surface
Planet's ozone layer	Water variation and timing of average rainfall
Planetary surface	Water absorption by planet's mantle
	Water vapor in atmosphere

35 of the 162 Constants Required for Life

Some theoretical scientists say that the Anthropic Principle arguments are poor. However, they do not offer reasons for disputing them.

The Odds Earth Was Formed by Random Events

With 162 constants identified in the Anthropic Principle, scientists have calculated the odds that a random event could have created each of the single constants.

Knowing the random event odds for each of the 162 constants, the total odds that Earth was created by random events was then calculated as one chance out of 10^{138} chances. That's one chance out of 1,000,000,000,000,000,000,000,000,000,000,000,0 00,000,000,000,000,000,000,000,000,000,000,000, 000,000,000,000,000,000,000,000,000,000,000,00 0,000,000,000, 000,000,000,000 chances.

The Anthropic Principal odds of happening randomly are astronomical. In comparison, the chance of winning the 33-State Powerball Lottery is one chance out of 10^8 chances, or one chance out of 195,000,000. Purchasing a lottery ticket is a *simple random event*, yet still having an incredible high probability of losing. Creating a habitable Earth involves successfully executing *complex creation events* one after another.

Was Earth created by random events? Given the odds, it seems unlikely. If Earth was created by random events was it by something totally unknown today and incomprehensible by scientists?

It is hard to imagine that a creator random event was so powerful and so smart that it could be responsible for all the creations in the past as well as all the creations that are still happening today.   Does this mean that the *creator random event* is still alive and well today?

Winning the lottery
10^8 or 1 in 195 million chances

Winning all NCAA brackets
10^{18} or 1 in 1,000,000,000, 000,000,000 chances

Can't happen rule
> 10^{50} chances

Random events created big bang
1 in $10^{9,221,000,000}$ chances

Random events created Earth
1 in 10^{138} chances

The Odds of Random Event Creations

Chapter 10

What Created Humans

Between the time of Earth's creation 4.5 billion years ago, and up until around 300 million years ago, there were relatively no living things on Earth. There were plenty of atoms but none inside the cells of living organisms.

A live atom is the same as a dead atom, only the live atom resides inside a living plant, animal, or organism. Dead atoms are still coming alive in cells, even today. How do dead atoms come alive?

Abiogenesis.

Abiogenesis

**The moment dead atoms
become living atoms inside a living organism**

The term abiogenesis refers to the moment that dead atoms become living atoms. Scientists have tried for years to make dead atoms come alive. They have never been successful.

Abiogenesis is an unexplainable creation step that transforms collections of dead atoms into living atoms inside a living cell. It only takes a few select atoms, namely carbon, oxygen and hydrogen, to become living atoms inside the cells of every plant and animal.

Dead atoms
Science laws, information, systems
Big bang, atoms, stars
Solar Systems, planets
Earth
Living atoms (abiogenesis)
Amino acids
Protein molecules
Single cells
Organs
Humans

Theoretical scientists say random events caused the big bang to be the creator of pieces of atoms followed by the assembly of complete atoms, and finally followed by the distribution of these atoms throughout the Universe. Atoms from long ago are still found on planet Earth and are still occupying the bodies of animals and plants.

Abiogenesis Gives Life to Atoms

Scientists avoid explaining exactly how the beginning of the evolution chain began. Because abiogenesis is so difficult to explain, scientists avoid this subject matter altogether.

Scientists have tried for years to make dead atoms come alive but have been unable to do so

When dead atoms become live atoms inside cells, something is added to the atoms enabling them to become a living thing. That something is called *information*. No one knows where the information is stored until it is needed or how it adds life to living cells. It just shows up when it's needed.

Scientists Cannot Create Living Cells

In the creation chain of events that transforms dead atoms into living atoms, *protein molecules* must be created for every cell of every specie, for every living plant and animal. Millions of protein molecules make up the composition of a single cell. You heard that right. A single cell. Remember humans are made up of 13 trillion cells. It takes a staggering number of different kinds of protein molecules to make up the cells of humans. Creation of the protein molecule was a major step for living things after Earth was created.

13.8 billion years ago
\
Systems, science laws, information
Big bang
Universe
Earth
Protein molecules
Single-celled animals
Humans
\
Today

Abiogenesis research has been going on for years, dating back to the 1950s. Several scientists proposed theories about life beginning in swamps. They claimed random molecules fell from the sky into swamp water and then were struck by lightning creating the beginning of life.

A single-celled Protozoa was the first living animal said to have been created. Scientists claim the Protozoa became instantly self-sufficient, living off bacteria in the water.

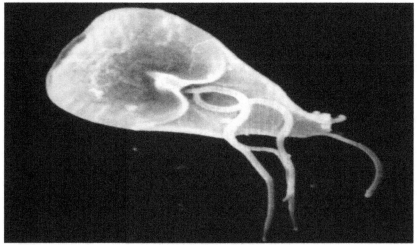

A single-Celled Protozoa

Creation of living single-celled animals and plants requires very sophisticated systems and information for the cell's molecules to become living and operational.

In conducting their abiogenesis research scientists filled a glass container with methane, ammonia, water vapor, and hydrogen gases. They theorized this represented the early atmosphere of Earth. Then they ignited a spark simulating lightning hoping to produce living molecules.

Because this experiment produced only amino acids, it partially worked. The experiment demonstrated that amino acids are relatively easy to produce. But it failed to produce living cells. Something was missing - abiogenesis.

The findings of the experiment showed that single-celled living organisms are extremely complex. Even if the test tube experiment had been successful in creating protein molecules, the next steps would have been even more difficult.

Early Atmosphere of Earth

The next steps would have involved creating living cells out of nothing, followed by creating internal *organs* like the heart, liver, lungs, pancreas, and brain, to name a few.

**Creating living cells is far more
complicated than scientists imagined
because during their creation
cells somehow receive information
and systems
from an unknown location
which makes them come alive**

Because of the complexities inside living cells, some scientists say the chance of a random event creating a single-celled living organism like a Protozoa is one chance in $10^{4,478,296}$ chances. This is extremely remote.

**The chance of random events creating
a single-celled Protozoa from dead atoms
is 1 chance in $10^{4,478,296}$ chances**

Another more complicated single-celled animal called a Bacterium resides in human's intestinal tract. It insures human survival as it breaks food down in the digestive system. Because of its increased complexity, some scientists say the chance that a Bacterium was created by a random event is one chance in $10^{100,000,000}$ chances.

Bringing atoms together to form different kinds of molecules is no small task. It requires a new set of systems and laws to hold the molecules together and manage them.

A *glucose molecule* called sugar, found naturally on Earth, is comprised of hydrogen, carbon, and oxygen atoms. Glucose is a good example of how creation complexity increases even for a simple molecule.

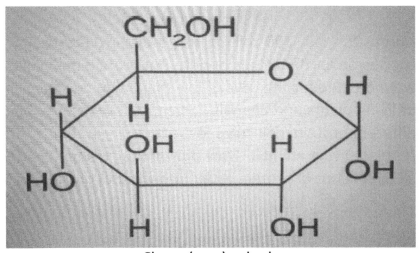

Glucose (sugar) molecule

Other molecules found in nature are far more complex. Molecules produced in laboratories are even more complicated with stringy structures that can fill an entire book page.

In the pursuit of new medicines, chemists alter atoms and molecules routinely in laboratories. People who have experienced positive health results from eating naturally grown plants for medicinal purposes sometimes offer clues to scientists. These plants can be molecularly synthesized to determine their chemical make-up and possibly produced as new medicines.

Did Random Events Create Protein Molecules?

A protein molecule is an extremely complicated essential component of all living cells.

During the last 300 million years of introducing new species on Earth, billions of different protein molecules were created for the cells of five million unique species of living plants and animals. Scientists say it takes 239 different protein molecules to create a single cell for the simplest living animal. How did random events create all the billions of protein molecules without one failure?

**It takes 239
different protein molecules
along with information and systems
to create
the simplest single-celled animal**

Hooking atoms together to form protein molecules is a huge task. Information as well as systems must be available prior to the creation of molecules for their successful creation and continued operation.

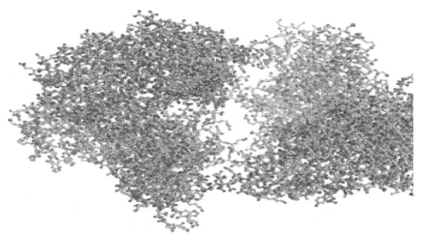

The Simplest Protein Molecule is Extremely Complex

According to some scientists the chance that a single protein molecule was designed and created by a random event is one chance in 10^{161} chances.

**The chance that random events
created the simplest protein molecule
is one chance in 10^{161} chances**

Did Random Events Create Single-Celled Animals?

The creation of the simplest living animal cell is the most complex creation event up to this time in the Universe's timeline. After exploring inside these single living cells and discovering their complexity, entire science research projects have been abandoned.

What the scientists found inside single cells were millions of intelligent activities, performing like perpetual motors running seamlessly. It was so complex scientists could not unravel or replicate them.

According to scientists, the simplest single-celled living animal contains 10^{80} atoms or 100,000,000,000,000, 000,000,000,000,000,000,000,000,000,000,000, 000,000,000,000,000,000,000,000,000,000 atoms.

Every atom in a human body is a self-contained perpetual running machine inside the molecules which are inside each of the 13 trillion cells. They all work together. The cell numbers required to make a human being function are staggering.

Systems Make Organs Come Alive

Creating living organs is also complicated. Organs are made of millions of their own specific cells and require their own specific systems and information to make them function.

Information and systems must be present before the living organs can be created. Systems of the body direct the operation of organs using communication signals and feedback. In humans there are 14 major systems governing the body. The blood circulatory system is an example of a system that is critical for keeping humans alive.

Without systems and information, cells are nothing more than dead atoms and molecules with no purpose or functionality. They would be like all the dead atoms on Earth lying around before life began. Systems and information were discovered by scientists and doctors. No one *created* them. Artificial intelligence programmers use systems and information routinely in their work.

Circulatory – heart, blood, and blood vessels
Urinary – kidneys, ureters, bladder, and urethra
Respiratory – nose, mouth, pharynx, larynx, trachea, bronchial tubes, and lungs
Digestive – mouth, pharynx. esophagus, stomach, liver, gallbladder, pancreas, small and large intestine
Nervous – brain, spinal cord, and nerves
Sensory – eyes, ears, nose, tongue, and skin
Endocrine – glandular or hormonal glands
Excretory – skin, lungs, liver, kidneys, and large intestine
Immune – lymphocytes and antibodies
Lymphatic – tonsils, thymus gland, liver, spleen, and lymph nodes
Muscular – muscles and tendons
Reproductive – testes, penis, ovaries, and uterus
Skeletal – bones, joints, ligaments and tendons
Dermal – skin, hair, and nails

14 Systems of the Body

Did Random Events Create Organs?

The systems and information required to operate a living cell cannot be changed by mankind because the location cannot be found. If the location of systems and information were known and could be manipulated, would it be the same as a computer program that is routinely altered? Would mankind dare to modify critical human system functions?

Systems and information appear to have been precisely pre-programmed long ago for living cells. They were stored somewhere ahead of time for cell creation and *available on demand* allowing cells to come alive.

Systems

A set of connected parts that use information, laws and reliance on each other to create a new complicated whole

---

There are many examples of systems and information working together, such as the pancreas and liver organs found in the digestive system. While each organ carries out specific functions, they also work together with other body organs.

The pancreas organ contains cells that secrete hormones to control the level of sugar in the blood stream.   These hormones are called insulin and glucagon.

When the body's blood sugar rises during a meal, the pancreas releases insulin hormones into the blood stream causing blood sugar levels to fall. Insulin also aids in opening body cells to accept the sugar necessary for energy.

If blood sugar levels are too low, the pancreas slows down the release of insulin and releases the hormone glucagon which signals the liver to release sugar. Sugar is released from the liver and enters the blood stream for cells to use as energy for the body when not eating.

The liver and pancreas organs work together (systems), to keep the body's blood sugar levels within a normal range between 70 and 120 mg/dl (information). This is a perfect example of systems and information working inside the human body. Under the control of systems and information, two separate organs are working together to keep the human body functioning.

How does this work? Where do the systems and information reside? There are hundreds of complicated systems like this in the body. Could random events create such remarkably complicated organs and systems?

Another remarkable feat in creation is the interaction of massive amounts of information between the nervous system and the sensory system. The eyes, nose, mouth, ears, and touch receptors all provide input data to the brain through nerves. How random events created a control system capable of directing the brain's nervous system, interacting with the body's sensory systems, has not been explained by scientists.

The human brain is the most complicated information processor in the Universe. It contains 100 billion cells. Each brain cell has 50,000 wires or neurons connecting it to other brain cells.

The brain cells have 120 trillion wires connecting them to all the cells throughout the body. This averages 1200 wires for each individual brain cell. Every second, the brain receives 100 million electronic signals traveling at 260 miles per second from each of the 13 trillion body cells back to the brain. Most people cannot comprehend this amazing feat. Did random events design this electronic sensory and cloud storage system?

Scientists say that if the human brain were to learn something every second, it would take three million years to exhaust the brain's storage capacity. Hard to imagine? How many yottabytes of storage is this? This is a storage capacity that dwarfs the largest computer processing centers. To think that this storage center is mobile and sits on top of human bodies wherever they go is amazing.

The brain is a central processor for all body activity and is very complicated. Its total capability is yet to be discovered. The eyes, nose, mouth, ears, and touch all transmit messages in parallel, simultaneously, to the brain. Depending on the type of message, the brain answers with commands that travel back to the source.

The eye is one of the most complex parts of the body. It is an unimaginable, intricate, well-executed design when working properly. If random events were involved with this creation, it would be considered its shining hour.

In visual recognition, the eye and brain determine the color, shape, depth, and movement of an object. Along the transmission path to the brain, billions of analog picture signals are converted to digital information packets that the brain can comprehend.

The eye is by far the most complicated analog to digital input conversion device known

To get information packets to the brain, they must swim through a liquid channel, called the synapse, which surrounds the brain. When these packets land on shore, they must arrive at the correct brain receptors, or docking stations to carry information to a predetermined location for storage and processing.

Mass retailers and artificial intelligence developers could learn something from this storage and retrieval system for use in their fulfillment centers.

In addition to separating the rapidly arriving external data sources coming from five sensors, the brain must store the information packets in an organized fashion. It must group together similar information to form an image. Later the packets may be called upon to remember something whenever the brain is in recall mode.

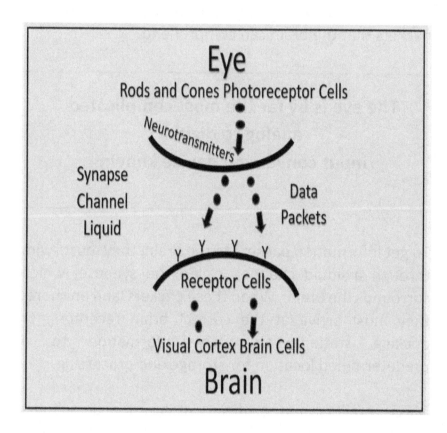

Two of the senses that rely on object and voice recognition are being studied by scientists with the hope of assisting humans who have difficulty functioning and for those who need to increase their productivity. Object and voice recognition are simulated in laboratories using newly-developed devices that employ perception, motor control and multisensory integration. These devices are being developed with the use of artificial intelligent software, neuromorphic engineering, and theoretical quantum computing using large scale computers.

Symbiotic Relationships

In the plant and animal world there are examples of *symbiotic relationships* that are taking place every day. Humans and animals exhale carbon dioxide as a waste product which becomes food for plant life. Plants convert carbon dioxide back into oxygen for humans and animals to breathe. Both activities keep each of the other living members alive. That's correct, they keep each other alive. This is an unimaginable symbiotic relationship example between plants and animals. How did random events, or aliens for that matter, create this sophisticated symbiotic relationship found in nature?

Symbiotic Relationships

When two living organisms
rely totally on each other to survive
either by choice or not

It doesn't matter whether you are living in the northern or southern hemisphere, the exchange of oxygen and carbon dioxide takes place seasonably and continually across continents with the help of the wind and the Earth's rotation.

There are many unexplainable examples of transfer of knowledge between living things. Some are unexplainable. This is acutely evident within the life cycle of Monarch butterflies. Monarch butterflies live only 90 days. Their short life span does not allow them enough time to know their parents or experience the migration route they must take. The most a single Monarch will experience is twenty-five percent of its migration journey. No one knows exactly how this happens, but butterflies are instinctively able to navigate migration routes between Mexico and Canada each year without any training.

On one segment of the Monarchs' journey, the mature butterflies migrate from northern Mexico across the Gulf of Mexico to southern US. They lay eggs on poisonous milk weed plants. Then the eggs hatch to form butterflies. As soon as these parents lay their eggs, they die. The new generation hatches and flies to northern US and Canada. The next generation flies back to the southern US, and the last generation migrates across the Gulf of Mexico back to northern Mexico, all journeys without any guidance from the Monarch's parents.

Other symbiotic examples between insects and plants, include flowers. Scientists have recently discovered electrical fields around flowers, allowing them to communicate with insects in a beneficial symbiotic relationship. Color, shape, and fragrances have always been thought to be the primary attraction for bees. This new magnetic connection of flowers with pollinators is a major benefit to both.

Is there intelligence in the design of relationships between plants and animals, or is everything *randomly* perfected? It seems there are too many plant-to-animal relationships beneficially designed for the survival of living things, to be randomly created.

Chimera – Passing Living Cells Between Hosts

When a mother carries a child in her womb, it is possible for her to pass her body cells directly to the baby through the placenta or through the mother's breast milk after child birth. It is also possible for cells of babies to pass cells back to their mothers and to other siblings while inside the womb. This is called *chimera*.

Chimera

The passing of living cells from a living host to another living host

Chimera also occurs when foreign body cells from donors are introduced after organ transplants. Transplanted body organ cells can live and function in another person for about ten years.

Stem cell research is starting to show progress in replacing damaged cells in the body by using stem cells harvested from young human bodies. The stem cells, when introduced into a patient, appear to be able to seek out damaged cells causing them to vigorously regrow back into cells as they were originally.

Apoptosis – Death to Living Cells

The opposite of abiogenesis is *apoptosis*. When a living cell dies, it releases its atoms and is called apoptosis.

Apoptosis

**Programmed cell death
releasing back into the Universe
the atoms, systems and information
that keep cells alive**

Apoptosis cell death can happen when living things die a natural death or because of premature causes before death. During a person's normal life, cells continually die either naturally or from injuries, but they are continually *re-created.*

When cells die, their protein molecules break down and the atoms inside return to Earth. Atoms are never destroyed. Atoms from dying plants or animals do not disappear. They remain on Earth and are redistributed. They are never destroyed. They are reused.

Cells also release to some unknown place the *information and systems* they relied on for survival.

**Atoms from dying plants or animals
are not destroyed.**

**They remain on Earth and are redistributed
and reused.**

Do Systems and Information Die at Death?

Are the systems and information in a cell reclaimed when they die, or do they disappear forever like a puff of smoke? Do they have to be re-created each time a new cell is created?

Adult humans re-create each of their 13 trillion cells every 3.4 years. When a person is thirty-four years old their body has been re-created ten times. This design must have been pre-programed to keep the body functioning longer.

**Every living cell in the entire human body
dies and is
re-created every 3.4 years**

Red blood cells in the circulatory system are re-created every 90 days. Glucose testing depends on the life cycle of these cells when averaging a three-month A1c blood test which is usually prescribed for patients with diabetes.

Are the systems and information, used to operate the dying cells, somehow transferred and reused by the newly-created cells through some process? Is informational memory, that took years to accumulate and has been stored in brain cells, transferred to new brain cells? If so, how?

Does Consciousness Die at Death?

For years materialistic scientists and psychologists have educated society to believe that the brain alone produces and retains *consciousness* inside the human body.

Consciousness **A person's awareness and responsiveness to themselves and their surroundings**

Consciousness is invisible. It is described as a person's awareness and responsiveness to themselves and their surroundings.

Theologians often equate *soul* to consciousness. Where the soul resides in the human body is usually not discussed from a theological viewpoint.

Soul

**The invisible core of a person
that is not limited by the body**

Until now, most scientists and philosophers have based their studies on what is referred to as a *materialistic world of reasoning*. This reasoning rejects all theories that a human body could function with consciousness outside the body.

If consciousness resides outside the body, how does it communicate with the body? For that matter, how does consciousness communicate even if it is inside the body?

Standard materialistic science theories have not allowed *non-materialistic evidence* into scientific work because it does not fit the definition of empirical science.

Does consciousness exist outside the body?

Today there are a few scientists studying *non-materialistic science* and theorizing that it is possible for consciousness to exist outside the body before and after the death of a human.

Materialistic Science

Research in which atoms and matter are the only reality

If this theory turns out to be a possibility, then an intriguing question remains - does consciousness reside outside the human body while a person is alive? This question will undoubtedly cause a pause in the scientific, philosophical, and theological worlds if this theory of consciousness is embraced as a possibility.

Non-materialistic researchers point to studies showing overwhelming evidence of people with *Near Death Experiences* (NDErs). NDE patients report having out-of-body journeys to different locations beyond Earth, and then coming back to life after being medically flat-lined from accidents, illnesses, or unsuccessful surgeries.

Doctors, documenting the stories of patients with NDE, consistently record patients' consciousness and journeys outside their bodies, although they were declared medically dead. Patients reported looking down over their own bodies during surgery, as they observed and heard their doctors in the operating room.

Over two thousand NDE events have been studied. The consensus reached so far is that it would be impossible for these patients to know about the people and places that they described in their journeys, after they were declared clinically dead.

Is there a consciousness that begins *before* creations and survives *beyond* death? Does consciousness exist outside a living body? If so, where does this consciousness come from and what is its creator?

The most complicated creation, of all known creations in the Universe are human beings. It is the final step in the creations of the Universe. Could a random event create a human? What would be the odds of doing so? How would a random event create the systems and information that controls body function? How would a random event design and create a consciousness?

Did Random Events Create Humans?

A theoretical cosmologist estimated the odds that humans evolved by chance due to random events at one chance in $10^{2,000,000,000}$ chances.

**The chance that that a human being
was created by random events is
1 chance out of $10^{2,000,000,000}$ chances**

When pulling together the odds of all the random event creation steps described in this book, is it possible that a *random event* creator could be responsible for creating the Universe?

Multiplying the odds of all the major serial creations together, the combined odds of humans' existence after the big bang are amazing.

Big bang   1 chance out of $10^{9,221,000,000}$ chances

Earth   1 chance out of 10^{138} chances

Protein   1 chance out of 10^{161} chances
  molecules

Protozoa   1 chance out of $10^{4,478,296}$ chances

Humans   1 chance out of $10^{2,000,000,000}$ chances

All Creations   1 Chance out of $10^{224,000,000,000,000,000,000,0000,000,000}$ chances

Is it possible, given the astronomical odds, that creations of the entire Universe were caused by random events (evolution)?

Creator Possibilities

Aliens

Random Events (Evolution)

God

Does this eliminate aliens and evolution as creator agents?

Part III

Chapter 11

Did God Create

Scientists say there are only three creation possibilities. For those who want to explore the possibility that God is the creator, because *alien* and *random events* (evolution) creation possibilities seem to be too remote, these last two chapters are about God, and all that is known about this creator possibility.

People with non-theistic views on God, gods, or an absence of a belief in any god, say that science is the sole creator in the natural world. They say religion's only role is to answer questions about the ultimate meaning of life and moral values.

One argument that some of the intelligentsia make is that the existence of God, as a creator, depends entirely on the proof of who created God.

Would this be like saying that the existence of evolution, as a creator, depends entirely on the prove of who created evolution?

In the 1980s, "creationist", "evolutionists" and "intelligent design" were terms used in creation discussions. Soon after, "theories" became the dominant science themes in US and world cultures. Books began to appear in classrooms and bookstores with science theories presented as facts. There were no disclaimers. Some classroom textbooks were approved without objective criticism from school boards.

Book content was complicated and difficult to accurately judge. Some college science professors were all too eager to reinforce *theories* as *facts* in public and legal debates. Reopening discussions of creation in public schools today, it seems, would be impossible because of intelligent design court decisions and the follow-on dominate opinion that evolution is the only possible creator agent.

If the study of *creation out of nothing* ever becomes a research subject of interest for scientists, will it be blocked because the courts have ruled the study of intelligent design cannot be funded in public institutions? Could a study like this be legally granted in public-funded universities?

Should physical science be expanded to include what *caused* the creations of the physical world?

Theories Dominate Science

Since the 1980s, science channels have portrayed celebrities as science experts explaining the latest scientific theories. Hollywood's influence can be a dramatic force for the adoption of theoretical stories as proof by eager audiences. Science magazines produce stories enticing science junkies to salivate at what *might* be true. Never mind, if it was a theory about an engine that could never run or a plane that could never fly.

Because of the digital-communication age today, the world at large is collectively getting smarter about science and creations. Many people are starting to question their intelligentsia. They want to know more than just theories. Some would like answers to the following:

- **What *created* the beginning of the evolution chain?**
- **Is there at least one genome evolution example of a new specie being *created*?**
- **How are living cells *created* from dead atoms?**
- **How were proton and neutron shells *created* during the big bang?**
- **How was the boson particle *created* out of nothing?**
- **How are simple protein molecules *created*?**
- **What *created* science laws, systems and the information they use?**

What if the Universe really was created by a supernatural creator, such as God? Public education of this idea is prohibited, and open discussion is discouraged. Non-empirical supernatural creators using "intelligent designs" were determined to be religious and illegal to be taught as creator agents in an opinion given by the US courts in the 1980s.

The argument in courts have been that creations by God could not be subjected to empirical testing. Has *evolution*, as a creator agent, ever been proven by empirical testing in the courts? No. Then why is evolution (random events) still being taught as if it is a creator agent?

Should the veil be removed that has been draped over *creations out of nothing,* opening-up the study to *all* the three possibilities for creator agents, not just the one chosen by scientists? Would scientists and theologians be open to this?

Does admitting that the beginning of the creation chain of events for things like the big bang, Earth, and living things, alter scientists' work? Hardly. It only admits that the beginnings of the design and creation chains really did have their origins in something supernatural, other than what scientists try to explain when using only material science.

A Design Plan

It seems likely creations of anything approaching the magnitude of the Universe requires a design. No human could have done it. Some people brush this topic aside because it appears to suggest a supreme being's involvement. As was stated earlier, designing without advanced planning, does not work because it does not provide a succession plan. A random event design plan would be too unreliable.

Should *where* a design plan came from be ignored, just because it cannot be explained?

Here are some lingering tough questions:

- **How could *creation* of a simple hydrogen atom be accomplished without some sort of a design?**

- **Without a design, what would any atom look like?**
- **How would an atom be built and from what would it be built?**
- **What would be the atom's intended purpose?**
- **Would an atom survive forever, or would it just wither away and die?**
- **What gave an atom its characteristics?**
- **What informs an atom to act as a perpetual machine and never stop?**

Creating something out of nothing is unimaginable. No one has figured it out, although science has tried. Scientific theories abound on how something "might be", or "could have", with no concrete evidence.

It seems far easier to ignore the truth about the complexity of creating something out of nothing, knowing that the intricate details lie far beyond mankind's abilities.

Is there a God?

In the 12th century BC the concept of one main god, as a religion, was introduced by people in South Asia. What caused them to believe this? This religion did not focus on salvation, but rather on mankind's actions, words, and deeds. In the Genesis book of the Bible, mankind was introduced to the concept of a single god around 2000 BC in the book of Job. According to the Genesis book in the Bible, Moses met with God on Mount Sanai and God revealed himself.

In the last chapter of this book (Chapter 12), the question of whether God, in human form, appeared to thousands of people on Earth, is discussed.

The Bible, or at least portions of it, is considered a religious book by Jews, Christians, and Islam. They say it was divinely inspired. It is filled with sub-books written by prophets, kings, and apostles between 2000 BC and 90 AD. This book has been translated into over 1400 different languages. The Bible is by far the most widely distributed book of all time with over 6 billion printed copies in circulation plus millions of electronic copies.

Some people question the authenticity of the Bible, because they believe it has been rewritten by mankind over the years. They are correct that the Bible has been changed in major ways over the years to fit a few religious leaders' interpretive differences, but for the most part it remains unchanged. Sometimes, whole new religious books are written to replace the original Bible, to better accommodate the religious leader's (intelligentsia) new-found religion.

Theoretical   **Provable**

Date	Event
?	Science laws and systems created
13.8 billion	Big Bang Universe created
4.5 billion	Earth, protein and bacteria molecules created
300 million	5 million species created
2.8 million	South Africa homo naledi
9,000 BC	Jericho continuously inhabited
2,500 BC	Papyrus & ink discovered
2,000 BC	Job book
1,500 BC	Moses book
800 BC	Homer book
700 BC	Isaiah book
50 BC	Thessalonians Book
59 BC	Livy book
65 BC	Mark Book
80 AD	Luke book
93 AD	Josephus book
1608 AD	Telescope
1665 AD	Calculus
1939 AD	Ramsay book
1948 AD	Carbon dating
1990 AD	Hubble telescope / rocket

Category groupings:

- Space; Atoms
- Gravity; Energy and matter; Stars and planets
- Carbon-dated writings; Historians
- Carbon-dated fossils; Archaeologists; Theoretical paleontologists and biologists
- Telescope; Calculus; Rockets; Theoretical physicists and cosmologists

Timetable of Living Things

Aside from changes to the Bible, people also argue that the Bible's dates cannot be authenticated and therefore the Bible's history cannot be proven accurate. Because the stories happened such a long time ago, there is no proof that, over time, the stories in the books were not rewritten by man.

There are many historical books written, in addition to the Bible, that authenticate the accuracy of events, including dates, places, and names of people depicted in the Bible. Some of these books include "The Book of the Kings of Judah and Israel", "The Records of Shemaiah the Prophet", "The Records of the Seers", and "The Commentaries on the Book of the Kings".

Some of the oldest cities in the Middle East are mentioned in the Bible, detailing stories about historical events. Two of those cities are Jericho and Damascus. These cities have been continuously inhabited for over 11,000 years. For the period the Bible covers, it is said to be one of the most accurate historical books ever written.

Jericho, West Bank	**9000 BC**
Damascus, Syria	**9000 BC**
Byblos, Lebanon	**5000 BC**
Sidon, Lebanon	**4000 BC**
Medinat, Egypt	**4000 BC**
Gaziantep, Turkey	**3650 BC**
Rayy, Iran	**3000 BC**
Jerusalem, Israel	**2800 BC**
Tyre, Lebanon	**2750 BC**
Kirkuk, Iran	**2600 BC**
Jaffa, Israel	**2000 BC**
Hebron, West Bank	**1500 BC**
Gaza City, Gaza	**1000 BC**

Oldest Continuously Inhabited Cities in the Middle East

In 1946, copies of Old Testament books were found by nomad Bedouin sheepherders in a cave at Qumran, next to the Middle East's Dead Sea. The arid climate preserved the parchment copies for over 2300 years.

Working with fragile broken pieces an international team of interpreters painstakingly pieced together a complete copy of the Isaiah book. Today a worldwide effort is ongoing to meticulously translate other books found at Qumran.

185

Qumran cave where the Isaiah book was found

Compared with today's Bible, scholars say the words in the Isaiah copy found at Qumran are 95 percent word-for-word the same as in today's Old Testament.

Fragments of a 2300-year old Isaiah book copy

The Bible is the only book that has copies dating as close in time to the actual date when the book was originally written. Other historical books have a longer gap in time between these dates than the Bible.

Book	Author	Date Written	Oldest Copy in Existence	Gap In Time	Copies In Existence
Isaiah	Isaiah	680 BC	300 BC	380 years	1
Iliad	Homer	750 BC	250 BC	500 years	643
History of Rome	Livy	59 BC	400 AD	459 years	1
Gallic Wars	Caesar	100 BC	900 AD	1000 years	10
Plato	Plato	400 BC	900 AD	1300 years	7
History	Thucydides	460 BC	900 AD	1360 years	8
History	Herodotus	480 BC	900 AD	1380 years	8

Oldest Known Copies of Books

Carbon Dating Authenticates the Bible

Throughout the last 4000 years of translating the Bible from one language to another, the Bibles in existence today have the same content as the oldest copies known.

The copy of the book written by Isaiah carbon dates to 300 BC with only a 380-year gap in time after the original book was written in 680 BC. In comparison, the book of Plato has a gap of 1300-years between the original book and the oldest known copy. With no one having original copies of either book, it is unlikely these books were rewritten during these early periods. Some people seem to question the authenticity of the Bible but do not question other historical books. No one questions whether the book of Plato was rewritten during this 1300-year gap.

Scholars declared the books of the Bible accurate in 189 AD. The complete Bible was introduced officially as one book in 325 AD. Also, in 325 AD the Roman Emperor Constantine converted to Christianity, declaring that Jesus and God were one. He called together the Council of Nicaea, comprised of bishops in the Roman empire (see Chapter 12). They were ordered to select books for the Old Testament and books written about Jesus to be included in now what is called the New Testament of the Bible.

Emperor Constantine

After this, Constantine asked his mother Helena to oversee the construction and preservation of holy sites associated with Jesus. She instructed governors to build 80 new churches in Bethlehem and Jerusalem including the Church of the Holy Sepulcher. This new church was a memorial of the crucifixion and burial sites of Jesus. According to the Bible, the actual burial tomb at Golgotha was provided by a rich Jew, while the crucifixion site was Skull Hill just outside Jerusalem's north wall.

Archeologists Authenticate the Bible

In the early 1900s, an archeologist used the book of Acts in the Bible, written by an apostle named Luke who was also a medical doctor, to search for seven lost cities that were named in the book. The cities were meeting places for early Christians fleeing Jerusalem after they were persecuted by the Romans. The seven lost cities were mostly in Turkey. After 2000 years of history, these cities were deserted and buried by dust.

By 1939, when the archeologist finished his work, he discovered that all the seven cities that were meticulously documented in the year 80 AD in the Book of Acts, were found exactly where Luke said they were. Archeology seems to confirm the authenticity of the Bible.

Where is God?

Moses wrote in the Genesis book that he saw God. Five hundred years earlier, it states in the Job book of the Bible that Job had conversations daily with God. Moses reported that Abraham talked with God as well.

God is mostly thought of as invisible with no physical form. Some words that are universally used to describe God include the following:

- **Unlimited power (omnipotent)**
- **Present everywhere (omnipresent)**
- **Perfectly good**
- **Loving**
- **Not needing material things**

One of the most difficult questions for people to answer is, where is this God? It seems most people do not think about or acknowledge God except perhaps at death or during difficult times.

For the scientists who want to see more proof, not being able to see God is a problem. Not being able to see the big bang does not appear to be a problem. Some scientists believe they are close to proving there is no God. They cannot prove, however, how creations out of nothing happened or provide empirical prove of the big bang.

Some theologians say God is impossible to see except in things that have been created. NDErs might argue this point because of their descriptions of God during their encounters with a super being.

To others, God is known only through his works. Seeing God through the marvel of what has been created and through continual creations seems enough for many of those with a religion and even for some without religion.

When scientists say there is no God and religious leaders cannot prove there is a God, few have answers. Those with curious minds continue to seek and ask questions.

How early in history, before Biblical accounts, did people know about God? Did some people see God as a visitor to Earth in a body that they did not recognize until later?

Is God the creator everywhere, all the time - even today?

Chapter 12

Where Is God

Five billion people on planet Earth believe there is at least one god. Some believe in multiple gods. They all believe their God or gods are omnipresent (present everywhere) and at the same time. Some believe their god lives everywhere in the Universe, and even within themselves.

The religions of Christianity and Judaism believe in one omnipresent God that is interwoven into everything, including all material science. The Islam religion believes in one God creator but separate from the material Universe.

Religions believing in multiple gods think their gods and the Universe are one but not limited by time or space inside or outside the Universe.

> **The religions of Christianity and Judaism
> believe in one God that is
> omnipresent and interwoven into everything
> including all material science.**
>
> **The Islam religion
> believes in one creator God who is
> separate from the material Universe.**

Four billion people on Earth believe their one God exists and is the designer as well as the lone creator of the Universe. For some of these people, it is not hard for them to believe God could appear on Earth at any time, looking and acting like an ordinary person, only with extraordinary powers and wisdom that no human could possess.

Did God visit Earth?

If God is the creator of the Universe, could God also go anywhere, do anything, be anything, anytime? There are compelling stories in history books, as well as the Bible, pointing to a visit by an extraordinary person in the Middle East 2,000 years ago who was like this. Today, billions of people are still talking about this amazing physical man that visited Earth. The Bible says his Earthly name was Jesus.

The Isaiah book in the Bible makes many specific *predictions* about God coming to Earth. Other books in the Bible like Matthew, Mark, Luke, and John were written between the years 50 AD and 90 AD and documented the story of God visiting Earth.

It is estimated at least 70 percent in the world have heard the name of this man Jesus who had supernatural powers and wisdom. In the US a Christmas holiday acknowledges his physical birth on Earth and is celebrated by not only Christians, but non-Christians as well.

Christmas Celebrated in the US

Christians	96 %
Non-religious	87 %
Buddhists	76 %
Hindus	73 %
Jews	32 %
Islam	0 %

Who is God?

People say God is invisible. Some say God can be seen in the unbelievable beauty of nature or the wrath of natural destruction. Others say the birth of a child or death of someone reveals divine creation plans from supernatural sources.

Another way to find out more about this invisible God is to look at what science has discovered while exploring creations. If God is the creator of the Universe, then scientists unknowingly are studying pieces of the creation puzzle that God created.

Other ways of discovering more about God is by reading religious books like the Bible. Some say the Bible is a product of man while others say it was devinely inspired. The Bible is a 4000 year old historical book, mostly dedicated to God. Accounts in the Bible are continually being studied today. The Bible says that God appeared on Earth in a human form, talking directly to people in the Middle East.

Some religious leaders do not accept Jesus as God, or the same as their god or gods, although they are aware that he made the claim that he was God. They believe Jesus was a great moral prophet and capable of doing amazing miracles that no mortal could do, but the religion that they follow does not allow them to accept God in human flesh walking around Earth.

Would an alien, walking around Earth, doing the same things as this man did, be rejected the same way?

Other religious leaders say Jesus was a lunatic, heretic, liar, and trouble maker. The Bible talks about him upsetting the religious laws that were established at that time and enforced by religious leaders. It also talks about Jesus becoming angry when religious leaders made a mockery of religious centers by turning them into marketplaces.

If he was a lunatic, how was he able as a young boy with no formal training and not then a threat to the religious leader's power, astonish religious leaders with knowledge far beyond his years, as recorded by historical scholars?

The Bible talks about this man being put to death without cause for being a troublemaker during a highly-agitated period in a country that was occupied and controlled by a foreign power.

Did this man undermine the intelligentsia's position of power and esteem during that period in history? Would the same thing happen today?

Most Religious Books Are About God

Many religions throughout the world have their own holy books which are about God. The Jewish Torah religious book contains five books that are the same as the first five books in the Old Testament of the Bible. There are 66 books in the Christian Bible with seven additional books added to the Catholic version. The Islam Quran contains modified versions of the Torah as well as modified versions of the Bible books of Psalms, Matthew, Mark, Luke, and John.

There are at least 20 different versions of the Quran holy book. There are 49 different versions of the English language Bible. One of them, the King James version, has had 2000 revisions since it was written in 1611. There are over 6900 languages in the world. To date, the Bible has been translated into 469 of those languages.

The Bible is a history book. Some say the 39 books in the Old Testament contain prophecies about God coming to Earth. The 27 books in the New Testament of the Bible tell about God's appearance as Jesus, a human in flesh on Earth, over a 33-year period.

Jesus Said He Was God

It is recorded in the Bible that this man, seemingly from another place other than Earth, but looking like someone from this Earth, said the following:

> **If you had known who I am then you would have known who my Father is**
>
> **From now on you know him and have seen him**
>
> **The words I say are not my own but my Father's words who lives in me and does his work through me**

Bible Book of John 14

When Jesus was asked by his followers, known as apostles, to show them who the creator was, he told them he was the creator known as God.

Historians Wrote About Him

Some of Jesus' worst enemies became historians later in their lives, recording Jesus as a man of extraordinary powers and wisdom. Others did not know him directly, but as part of their paid work as historians, recorded stories about Jesus as well.

In 93 AD a Roman history book, called *Antiquities of the Jews,* was written by a Jewish scholar named Flavius Josephus and included stories about Jesus. Through his writings, Josephus authenticated Jesus' time on Earth in the Bible. He did his research from interviews with Jews, Romans, and Gentiles.

The Roman government conscripted Josephus to write this history book because of his accurate writing skills. The Romans always documented their history, especially when occupying foreign countries.

Josephus was a lifelong Jew and had both Roman and Jewish citizenships. In his early career, he was a Jewish general fighting against the Roman army until he defected to advise the Roman legion. This new role labeled him a Jewish traitor.

To this day Josephus' historical work is shunned by most Jewish people. However, Josephus is still considered one of the best historians of his time.

Because Josepheus remained a lifelong Jew, it is highly unlikely that he needed to include the words that he wrote in his history book, about the man called Jesus.

Historian Flavius Josephus

His writings mention Jesus and his crucifixtion. Josephus called him the Messiah. He also said that Jesus was indeed an extraordinary man, "if indeed he was a man". He also included in his book the names and stories that were the same as names and stories in the Bible including Pontius Pilot, John the Baptist as well as James. Josephus' writings have drawn acclaim as an accurate and objective history book, based on his many interviews with a diverse cross-section of people.

- **At the time there appeared Jesus a wise man if indeed one should call him a man for he was a doer of startling deeds and a teacher of the people who received the truth with pleasure**
- **And when Pilot, because of an accusation made by the leading men among us, condemned him to the cross, those who had loved him previously did not cease to do so**
- **For he appeared to them on the third day living again just as the divine prophets had spoken of these and other wondrous things about him**
- **And he gained a following among many Jews and among many of Greek origin**
- **He was the Messiah**
- **And up until this very day, the tribe of Christians named after him has not died out**

Another historian, a Jewish man by the name of Saul, lived at the time of Jesus and wrote extensively about him between 50 and 90 AD. He was born in Tarsus, Turkey but lived in Jerusalem. He was the son of a high-ranking Jewish Pharisee leader. Saul, a well-educated man, hated Jesus and his followers. Saul felt the apostles' movement distracted the Jewish religious leaders and their efforts to appease the Roman government who took over their territory in 63 BC.

After the Romans increased their powerful stranglehold over the Jewish citizens, Saul made it his personal goal to wipe out the disruptive Christian followers who remained in Jerusalem after Jesus' crucifixion.

With the intent on killing all Christians, Saul took a band of Jewish guards with him to chase the Christians on their way to Damascus, after they fled Jerusalem. The Bible records that on Saul's way to Damascus he was stopped by Jesus and struck by a bright light which blinded him. This was witnessed by the Jewish guards accompanying him. Saul was blinded for three days and visibly overwhelmed by his experience. He suffered with his blindness until he obeyed the instructions given him by Jesus. After obeying Jesus' instructions, Saul miraculously regained his eyesight.

Paul Said Jesus Was God's Image on Earth

It was an epiphany for Saul when he saw Jesus alive again, after previously witnessing his death. Some say Saul's encounter with Jesus, was the same as a near death experience (NDE). Because of this life-changing event, Saul was transformed from a hater of Jesus to one of his strongest supporters and changed his name from Saul to Paul.

In 62 AD, Paul wrote to church members at Colossae Turkey, telling them that when the apostles were with Jesus, they saw *the physical image of God on Earth.*

Because Paul had both Jewish and Roman citizenship, he ministered to both. Paul's devotion to this extraordinary man called Jesus, lasted until the end of Paul's life. He is credited with writing many of the Bible books in the New Testament except the books of the four Gospels called Mathew, Mark, Luke, and John.

Paul was ultimately called the Apostle to the Gentiles. The Jewish religious leaders, of whom Paul had previously been a member, rejected him. Paul, as a follower of Jesus, shaped the future of the early Christian church more than any of the other apostles.

The story of Paul, as written in history books, is about someone who purposely removed himself from a lucrative Jewish life and member of the Jewish Pharisee religious council. Because of this he was banned from his Jewish religion and lived out the rest of his life as a poor dishonored follower of Jesus. The same can be said of all the eleven apostles who were persecuted for the rest of their lives.

After Jesus' Crucifixion He Was Called *God Jesus*

In 110 AD an Italian author and administrator named Pliny the Younger sent a letter to his boss, the Roman Emperor Trajan, asking advice on whether he should continue slaughtering early Christians, stating that their only crime was worshiping *God Jesus*.

In 178 AD a Greek philosopher, pagan opponent of Christianity and self-appointed intelligentsia named Celsus, published writings referencing Jesus as having great magical powers and calling himself *God*.

An archaeologist discovered the remains of a Christian prayer hall, carbon dated to 230 AD, 20 miles south of Nazareth in the ancient Jewish-Samaritan village of Othnay. This discovery represents important historical evidence that 100 years *before* the Roman Emperor Constantine embraced Christianity, Christians were already worshiping *God Jesus* appearing on Earth.

Critics of Christianity claimed the belief that Jesus was God did not start until *after* Emperor Constantine and the Council of Nicaea made up the story and declared it to be so in 325 AD. History proved this to be untrue.

> ## Critics say the claim
> ## Jesus was God (God Jesus)
> ## was made-up by
> ## the Council of Nicaea in 325 AD

A mosaic of *God Jesus* Christ has been uncovered on the floor of this newly discovered Christian prayer hall, calling Jesus a deity, confirming the New Testament's record that *Jesus was worshiped as God by early Christians*. It also confirmed that Jesus was continually worshiped as *God Jesus* after his death, until and after the time of the Council of Nicaea in 325 AD.

The building housing the Christian prayer hall was a residential building, measuring 30 by 40 meters, with a typical Roman-tiled roof, distinct from local-residents who used no tile. Somehow, Christians were able to use the Roman soldiers' living quarters and convert their common eating room into a prayer hall.

A mosaic was imbedded into the floor as well as a Eucharist table installed for their Last Supper communion. Does this suggest friendly interaction between early Christians and the Roman Empire, instead of persecution, which had been previously thought? The mosaic on the prayer hall floor was 54-square meters (580 square feet) with a Greek inscription stating that: "The God-loving Akeptous has offered the table to *God Jesus* Christ as a memorial." Scholars speculate that Akeptous was a woman who donated a table to be used for their communion ceremony.

Another inscription mentions a Roman officer named Gaianus who donated his own money to have the mosaic made. The mosaic also included fish, which was an early Christian symbol.

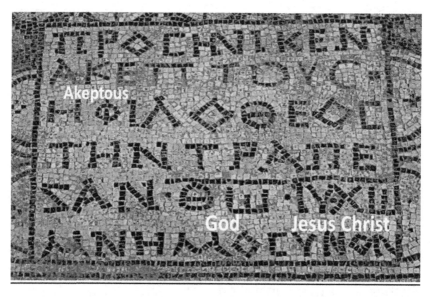

The God-loving Akeptous has offered the table to

***God Jesus* Christ as a memorial**

Skeptics Declare *God Jesus* Was Made Up

In 2003, a fictional book sold over 80,000,000 copies declaring the Biblical story that *Jesus was God* was fiction. The book stated:

(1) the Council of Nicaea made-up the idea that Jesus was God

(2) the bishop's vote barely passed by only two votes

(3) the Emperor Constantine banned, burned and had the original Gospels re-written

(4) the Vatican was the benefactor of Constantine's proactive role in presiding over the council

Historians have found none of the fictional book's claims to be true. Instead they found:

(1) discussions at the Council of Nicaea were not about *whether* Jesus was God, but rather how to convey to pagans *how* God could possibly appear on Earth as a mortal. Some philosophers, as well as the Islam and Jehovah's Witness religions, still use this reasoning as the basis for rejecting this claim. Some even regard as blasphemous the idea that God, if he wanted to, could turn into a

mortal with human weaknesses and servitude. Do people try to shape God into a human image?

(2) only two of the 250 bishops attending voted *against* ratification

(3) Constantine did not participate in the Council meetings and had no predetermined agenda, other than religious unity with pagans. The four Gospels used by the Council were carbon-dated to 200 AD, were never changed at the council and are readily available to evaluate today.

(4) the Pope did not attend the eight-week council. He sent only six Roman Catholic bishops from Italy. Out of the 250 bishops attending, the rest were from Eastern Orthodox religions in Egypt, Greece, Palestine, Syria and Turkey.

The Council of Nicaea was not about
***whether* Jesus was God**
but rather how to convey to pagans
***how* it was possible for God to**
appear on Earth as a mortal

Two-hundred years later, in one of religion's weakest hours of the 6th century, *power, privilege, deceit and command for respect* replaced the 1st century Christian church's establishment of *love, service and humble persuasion* as the foundation for its church.

Is it possible, when Jesus came to Earth, that a "picture" of God was left for those who ask what does God look like? Today, some theologians emphasize that Jesus was God in a physical body while on Earth. Most do not. Are todays Christian followers confused more than early Christians?

**Did God leave us a "picture"
of himself as Paul declared
when Jesus appeared on Earth?**

The Bible Told of His Coming

There were over 300 predictions, or prophecies, recorded in the Old Testament of the Bible between 1400 BC and 518 BC, about the coming of a great ruler. All of them were fulfilled.

The prophecies predicted this ruler would come to help the oppressed people of Israel. Most thought this meant saving them from the harsh leaders of Rome. The New Testament in the Bible declared the prophecies were fulfilled when God appeared in the middle east as a human, not as a ruler of their people, but rather to save people from an eternal death, giving them everlasting life.

Some of the prophecies recorded in the Old Testament of the Bible predicted how this man would be treated and how he would die:

- **There will be nothing beautiful or majestic about his appearance that will attract us to him**
- **He will be a despised and rejected man full of sorrows and bitter grief, but we will not care**
- **We will turn our backs on him and look the other way when he goes by**
- **He will be led like a lamb to slaughter and be silent to the shearers**
- **He will have done nothing wrong or never deceived anyone yet will be murdered like a criminal, but somehow buried in a rich man's grave**

Prophecies

The prophecies also declared God would be disappointed when he appeared on Earth because most people would have little understanding of who he was, of his teachings, or his Earthly mission.

The question of whether a seemingly human person could accomplish these prophecies has often been asked. In response to this curiosity, individual prophecies were extensively examined in college statistics classrooms to determine the odds that a mortal could have executed these biblical prophecies.

Their resulting calculations showed that the odds would have been extremely remote that a human could have accomplished even a few of these prophecies, yet alone 300 of the total prophecies written about him in the Bible.

Out of 300 prophecies, there were 30 specific prophecies in the Old Testament that Jesus fulfilled in just the last 24 hours before his crucifixion. Most of these prophecies were found in the Old Testament books of Psalms and Isaiah.

The prophecy fulfilments were all serial events. None of the events were parallel events. No one could have completed 300 parallel prophesied events at the same time.

Serial	Event	Prophesied	Fulfilled	Odds one chance in
1	Born in Bethlehem	Micah 5:2	Matthew 2:1	10^5 chances
2	Preceded by a messenger	Isaiah 40:3	Matthew 3:1	10^3 chances
3	Rides on a donkey	Zechariah 9:9	Matthew 6:11	10^2 chances
4	Betrayed by a friend	Psalm 41:9	Matthew 10:4	10^3 chances
5	Sold for 30 pieces silver	Zechariah 11:12	Matthew 26:15	10^3 chances
6	Price of potter's field	Zechariah 11:13	Matthew 27:5	10^5 chances
7	False witnesses	Psalm 35:11	Matthew 26:59	10^3 chances
8	Executed with thieves	Psalm 22:16	Zechariah 12:10	10^4 chances
	Forsaken by disciples	Zechariah 13:7	Matthew 14:50	
	Silent before accusers	Isaiah 53:7	Matthew 27:12	
	Cursed and spit upon	Isaiah 50:6	Matthew 26:57	
	Fell under his cross	Psalm 109:24	John 19:17	
	Hands and feet pierced	Psalm 22:16	Luke 23:33	
	Hated without reason	Psalm 69:4	John 15:25	
	Lots cast for his garment	Psalm 22:18	John 19:23	
	His bones not broken	Psalm 34:20	John 19:33	
	He was heartbroken	Psalm 22:14	John 19:34	

Odds of Completing 8 of the 300 Prophecies

By calculating just 8 of the 300 serial prophesied events, the odds that a single human could accomplish them on their own turned out to be one chance in $10^{16,200}$ chances. The odds seem astronomical that 300 prophecies could be fulfilled by a single human mortal.

The odds that anyone could have successfully executed only 8 of the 300 prophecies is 1 chance out of $10^{16,200}$ chances

After Jesus' crucifixion, the Bible reports at least 500 witnesses saw and spoke with him when he reappeared on Earth. It is recorded he appeared instantly inside locked rooms speaking and sharing meals with his apostles just as he said he would. The New Testament of the Bible says he reappeared over a period of forty days before he departed.

After he left Earth, it is said his apostles began to understand more clearly what his words meant. The prophecies about him, his life on Earth and why he came to Earth had been revealed to the apostles before their very eyes. They were first hand witnesses.

Miracles Recorded in the Bible

The Bible records that Jesus astonished people with many miracles. Yet they did not believe he was God in human form. Jesus told his apostles that most people were only interested in seeing his miracles and not interested in listening to his message.

**Most Christians and Muslims
believe the supernatural miracles
performed by Jesus really happened.**

Jews and skeptics do not.

Turned water into wine	John 2
Saved worker's son from dying	John 4
Healed a lame man by pool	John 5
Healed apostle's mother-in-law	Matthew 8
Provided large catch of fish	Luke 5
Healed man with leprosy	Mark 1
Healed paralyzed man	Matthew 9
Healed man's hand on Sabbath	Mark 3
Brought widow's son back to life	Luke 7
Calmed sea of Galilee	Matthew 8
Sent demons into herd of pigs	Mark 5
Healed bleeding woman	Luke 8
Brought girl back to life	Luke 8
Healed the blind and mute	Matthew 9
Walked on top of water	Mark 6
Healed all who touched him	Matthew 14
Sent demon out of girl	Mark 7
Healed many people	Matthew 15
Restored sight to blind man	Mark 8
Transformed on mountain	Luke 9
Sent demon out of boy	Matthew 17
Healed 10 men with leprosy	Luke 17
Healed blind beggar	Mark 10
Fed 5000 at Sea of Galilee	John 6
Healed man blind from birth	John 9
Brought Lazarus back to life	John 11
Rose from the dead	John 20

Miracles

Tell the Truth and Die – Tell a Lie and Live

Why would anyone voluntarily offer their life for certain execution? What would anyone gain except death?

The Bible indicates Jesus was not motivated by money, fame, prestige, or power to do what he did and said. One of the prophecies was about him handing over his life to his accusers. This prophecy was fulfilled when he was crucified.

The authors of the New Testament also had nothing to gain by telling the truth as they talked about and documented Jesus' life.

Getting nine different writers to report comparable stories while separated from each other is considered impossible, especially after long gaps in time between his death and when their stories were recorded.

The twelve apostles were paid no money, were persecuted, died poor, and yet did not deviate from telling the same consistent story about Jesus and what they saw with their own eyes.

Why were the apostles compelled to tell stories they had witnessed about his life and death? It seems the turning point for their becoming adamant followers was not until after Jesus' crucifixtion and reappearance after his death. Shortly after the crucifixion the apostles fled Jerusalem because they feared for their lives knowing they would be arrested whenever they spoke of his name and reappearance.

When Jesus reappeared alive to the apostles, they became unshakable followers for the remaining years of their lives. It has been said the crucifixion and the forty days of his visible reappearance on Earth marks the informal birth of Christianity.

The apostles had been with Jesus for at least three years, witnessing all his miracles and his crucifixion. It is recorded in history that the apostles, as well as thousands of non-Jewish people known as Gentiles, heard him speak with authority like no one they had ever heard.

Although Jesus was born into the Jewish religion, the Isaiah book prophesied he would be a light to the Gentiles, or non-Jews. He amazed the Jewish religious leaders with his vast unschooled knowledge of the Torah, especially when he taught as a mere child in the synagogue.

Before the apostles' brutal deaths, they authored all the New Testament books of the Bible between the years 50 AD and 90 AD. For merely telling their stories, all of them were killed as if they were criminals.

Why didn't at least one of the apostle's try to save their life simply by telling a lie, denying the stories about Jesus? None of them did. If someone is threatened by death, it seems most people would lie to save their own life. The apostles must have seen extraordinary events that convinced them to continue telling the story about God Jesus, even knowing they faced certain death. All of the apostles except John, who was exiled to a remote island, were brutally executed because they would not change their stories under penalty of death.

- **Stephen was stoned to death**
- **Philip was crucified**
- **Matthew suffered death by the sword**
- **James was beaten and stoned to death**
- **Matthias was beheaded in Jerusalem**
- **Andrew was crucified in Edessa**
- **Mark was dragged to pieces in Alexandria**
- **Peter was crucified in Rome**
- **Paul was executed by the sword**
- **Jude was crucified in Edessa**
- **Bartholomew was crucified in India**
- **Thomas was killed by a spear in India**
- **Luke was hanged in Greece**
- **Barnabas was killed by unknown means**
- **Simon was crucified**
- **John escaped death but banished to Patmos**

The Body in the Tomb Vanished

History records that Jesus was crucified and buried in a tomb in full view of hundreds of Jews, Gentiles, and a legion of Roman soldiers.

After he was crucified and died on a cross, Jesus' body was placed in a tomb donated by a rich Jewish man. The tomb was continually guarded by Roman soldiers.

An unimaginable prophecy was fulfilled, when after three days, his guarded tomb was found empty and Jesus reappeared to his apostles.

The Bible records that religious leaders paid Roman soldiers to lie about the empty tomb saying that the body was stolen by Jesus' followers. Some say it is illogical that his followers would be brave enough to steal a body out from under a large Roman legion of soldiers guarding Jesus' tomb, knowing the tomb was sealed. They knew the Jewish leaders were prepared to turn troublemakers over to the Romans for mandatory crucifixion.

Many have questioned how it would have been possible for a handful of unarmed followers to physically roll away a multi-ton boulder while a Roman legion of soldiers were guarding the tomb's entrance. Why would the Romans let anyone take the body away, especially after it had a Roman seal attached to the door, so no one could unknowingly enter the tomb?

Also, it is said that if the body was in the tomb, the Roman crucifiers would have been all too willing to ceremoniously bring Jesus' dead body out for public display as an example of what would happen to troublemakers. History books report that rotting bodies on crosses were often left on display for long periods of time to reinforce to people the price troublemakers would pay.

The apostles refused to back down and deny their claims that this human form of God on Earth was their Messiah, even though this Messiah had come to Earth to serve a different purpose than the Jews had hoped.

Is God Everywhere?

There are no other historical stories of a single person that have been perpetuated for so many centuries with such a high level of intensity as that of this man appearing on Earth. Maybe this is because of the following:

- **His story fulfilled the unbelievable odds of over 300 prophesies**
- **His miracles of healing and bringing people back to life**
- **His rising from the dead and being seen by over 500 people**
- **His compelling messages of love, forgiveness and life after death**
- **His followers' willingness to die instead of lying to save their life**
- **His crucifiers inability to parade his dead body proving him dead because it was nowhere to be found**

Did Mankind See God 2000 Years Ago?

Could Jesus have been God appearing in a human form? He performed many miracles of healing the sick and raising the dead before thousands of witnesses, which no one else could do. Hundreds of people saw him hanging on a cross and alive again after his crucifixion.

**Did people actually see God
2000 years ago
in a human form?**

If God came to Earth as a man and did what is reported to have happened in the Bible, then was he God appearing in a human form?

What's Causing Creations? Not likely that it was aliens. Not likely that it was random events (evolution). Is this mystical, powerful non-scientific possibility of God, as creator of something out-of-nothing, the only creation option remaining?

You decide.

Acknowledgments

Bruce Alberts
Andy Albrecht
Eben Alexander
Muqtada al-Sadr
Reza Aslan
Flavius Augustus
Charles Babbage
John Baldwin
Mara bar Sarapion
John Barrow
Joseph Bates
David Beckham
Isaiah ben Amoz
Nicodemus ben Gurion
Paul Benioff
Lee Berger
Jorge Bergoglio
Emile Borel
Dan Brown
Gautama Buddha
Bruce Buff
Warren Buffett
Paul Butler
James Cameron
Brandon Carter
Ernst Chan
Gilbert Chesterton

James Coppedge
Lee Cronin
John Dalton
Charles Darwin
Paul Davies
Richard Dawkins
Anuokpitoc Democratus
Michael Denton
David Deutsch
James Dobson
Lhamo Dondrub
J Presper Eckert
Mary Eddy
Albert Einstein
Richard Feynman
Steven Furtick
Pauline Gagnon
John Dixon
Martin Gaskell
Rick Gates
Bill Gates
Robert Goddard
Stephen Gould
Bruce Greyson
Johnathon Gruber
John Hagee

Acknowledgments (continued)

Ken Hamm

Steven Hawking

Richard Henry

Peter Higgs

Gordon Hinkley

Fred Hoyle

Ron Hubbard

Edwin Hubble

Peter Hughes

Muhammad ibn Abdullah

Abraham Ibn Ezra

David Jerimiah

Phillip Johnson

Flavius Josephus

Larry Kiser

Arthur Keith

Tracy Kidder

Andrew Kohut

Lawrence Krauss

Robert Lanza

John Lennox

Willard Libby

Hans Lippershey

Martin Luther

Joao Magueijo

Yuri Manin

Jeff Manion

John Mauchly

Carver Mead

Stanley Miller

Harold Morowitz

Thomas Nagel

Isaac Newton

Billy Nye

Irenaeus of Anatolia

Luke of Antioch

Moses of Goshen

Jesus of Nazareth

Saul of Tarsus

Mazlan Othman

Colin Patterson

Arno Penzias

Max Planck

John Polkinghorne

Karl Popper

William Ramsey

Joseph Ratzinger

Robert Reich

Alberta Rivera

Hugh Ross

Acknowledgments (continued)

Rick Rule
Larry Sanger
Gary Schwartz
Ralph Seelke
Joseph Smith
Richard Stallman
Columba Stewart
Peter Stoner
Lee Strobel
Mark Stoeckle
David Thaler
Steven Taylor
Nikola Tesla
Yotam Tepper
Peter Thiel
Frank Tipler
Noah Turner
Desmund Tutu
Harold Urey
Steven Vogt
Werner von Braun
Eugene Walcott
Jimmy Wales
George Ward
Rick Warren
Dawn Wendzel

Eugene Wigner
John Wycliffe

About the Author

DOUG HANS is a theoretical author who is not a scientist, theologian, or member of any organization - just someone like many others who are in awe of creations. Because he had more questions than science or theology could answer, he took a six-year research journey to write "What's Causing Creations", reporting on his research about creations by aliens, random events, or God. Early in his career he was inspired by Pulitzer Prize author Tracy Kidder's book "The Soul of a New Machine". He immediately saw a corollary in the design and operation of computers with the Universe.

He spent most of his career with international Fortune 500 companies in the management and consulting of engineering computer systems ushering in new technologies. As an indie filmmaker and author, he has produced documentaries and written a full-length screen play, about which he plans to write his second book. The author is a highly-educated graduate of a major university, with an analytical mind gravitating towards emerging science concepts.